EVOLUTION EXPLAINED

EVOLUTION
EXPLAINED

Peter Hutchinson

DAVID & CHARLES

NEWTON ABBOT   LONDON

NORTH POMPRET (VT)   VANCOUVER

575
H977

ISBN 0 7153 6190 2

Library of Congress Catalog Card Number 74-81008
© Peter Hutchinson 1974

Set in 11 on 13pt Baskerville
Printed in Great Britain by
Redwood Burn Limited Trowbridge & Esher
for David & Charles (Holdings) Limited
South Devon House Newton Abbot Devon

Published in the United States of America
by David & Charles Inc North Pomfret
Vermont 05053 USA

Published in Canada by Douglas David &
Charles Limited 3645 McKechnie Drive
West Vancouver BC

For F.L.B.

Contents

List of Plates

List of Plates

Preface

This book is intended as an introduction to the theory of evolution, and has been written for readers with no previous knowledge of the subject. If, by chance, it is read by more advanced students of biology, I hope that they will forgive the many simplifications and omissions in the text. I am well aware of these, but I am also certain that a book such as this loses much of its value if the author strays too far from his chosen theme by discussion of minutiae. I have not included references in the text, for they seem superfluous in an elementary work of this kind.

Frank Buckell and Dr Michael Dewar have read the text and their suggestions have been critical and helpful; I am grateful to them both for their encouragement. I would also like to thank Mrs M. Owens for producing the typescript from an almost illegible manuscript.

P.H.

I

Structural and Dynamic Elements

Biology is an extremely complex science. Its raw material consists of some 1,120,000 different living animals and 360,000 different plants. Many more are now extinct. Animals vary in size from those individuals that can live their entire lives within a single drop of water, to creatures such as a dinosaur called *Diplodocus* that was over eighty feet (twenty-four metres) long. In spite of such diversity, a single theory unifies all that we know about animals and plants. This theory is called the theory of evolution, and it states that animals and plants have changed since their origins several thousand million years ago. During the latter half of the last century, the task of biologists was to find evidence in support of this theory. They succeeded brilliantly and, today, research is directed at solving problems posed by the supposed mechanics of evolutionary change.

I want to begin by explaining how it is that a single theory can be so important to our understanding of a subject so complex as biology and, because I do not assume that the readers of this book have any knowledge of biology, I will start by talking about towns and cities, for they are similar in many ways to animals and plants.

Fig 1 shows the plan of a city. Although simplified, the plan shows that a city is extremely complex. How do we begin

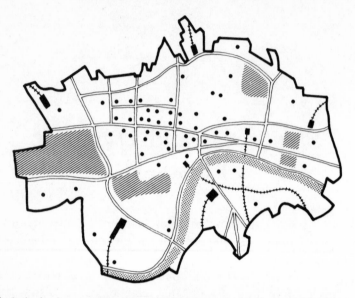

Fig 1 A city is a complex of structural elements such as roads, railways, and rivers. Each element has unique characteristics and functions, but all are integrated so that the city functions as a unit

to understand it? How can we investigate its structure, and how can we learn what happens within its boundaries? Well, to begin with, the structure of a city can be broken down into a number of items that are more amenable to investigation. This has been done in fig 2, in which a few of the structural elements of a city are isolated from other elements. Once this has been done, it is possible to think more clearly about the structure and function of each element. Roads are seen to form a network enabling people and things to reach every part of the city, whilst railways, such as those illustrated in fig 2C, are obviously concerned with the establishment of communications with points outside the city boundary. Because each structural element performs a function that cannot be performed as well by any other element, each has its own characteristics and modes of operation.

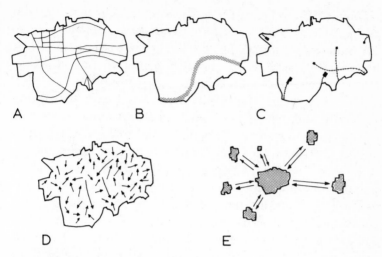

Fig 2 The elements of a city are of two kinds, structural and dynamic. Structural elements such as roads A, a river B, and railways C, have well-defined physical characteristics. Dynamic elements are less concrete although they may be closely related to structural elements. Dynamic elements are either internal D, operating within the city boundary, or external E, operating between a city and other communities

Another point becomes apparent when we think again of the city as a whole. No element, however efficient in itself, is really efficient unless it is integrated with other systems. A railway station is useless, for example, unless it is connected to other parts of the city by roads.

As well as structural elements, there are what may be called dynamic elements within a city. These could not be indicated in fig 1 because they have no permanence. An example of a dynamic element is the flow of money. Every day, large numbers of transactions take place in a city. They involve the transfer of capital, and may range from the purchase of a bag of sweets to the selling of a company. Such dynamic elements are indicated in fig 2D, and it can be seen that, although vital to the life of the city, such elements need not necessarily bear obvious relationships to the city's structural elements. As well as internal dynamic relationships, there are many relationships that are external (fig 2E). Using the example of capital flow once again, it is obvious that money is transferred between any one city and neighbouring communities.

It is clear then, that a city is a complex of interacting structural and dynamic elements. The next question we may ask is, how does such a complex come to be formed? The answer is obvious, for it is a commonplace that Rome was not built in a day. A city begins life as a village. Slowly, as its population grows, embryonic structures become more elaborate and, at the same time, dynamic aspects of the growing community become increasingly complex (fig 3). One could, of course, imagine a city that is planned in advance and then built (Brasilia is an example), but even cities such as these continue to develop after they are 'completed'. As a result of having grown, all cities have a past, and there is always evidence of the past in their structure. Early phases of a city's development can be identified by architectural styles of previous decades; also, parts of its fabric become obsolete but may be preserved for no other reason than that they are evidence of the past. Often it is possible to work out the order in which events occurred. If a railway, for example, crosses a road by a bridge, it is certain that the railway was built after the road.

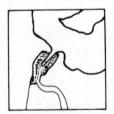   

Fig 3 The evolution of a city (Amsterdam from 1400 to 1900). Cities are not created *in toto*. As they evolve, their structural and dynamic elements grow in complexity

Very recent changes in a city may be deduced by comparing one city with other towns and cities. Hence if a building of the kind that in most towns functions as a cinema is found to contain scrap metal, it is reasonable to infer that the building's function as a warehouse is of recent origin and that it had previously functioned as a cinema.

At every stage of its development, a city is efficient, that is, it carries out certain functions, such as housing people and acting as a centre for their activities. Over a long period of time, however, the character of a city may change although it continues to carry out the same basic functions. Different subsidiary functions may produce cities which have different characteristics. Hence, a city on the coast that depends upon a port for its livelihood is different from a city located inland, although the two may have many structural and dynamic elements in common. Occasionally, the establishment of a new industry may completely change the character of a city in a remarkably short space of time. Conversely, some change may cause a city to cease functioning altogether. Such a change may be catastrophic (the destruction of Pompeii), or gradual (the decline of Rome as the centre of an empire).

We have seen that cities are complexes of structural and dynamic elements, and that these elements are closely integrated to enable the city to carry out its basic functions. Such complexes can only occur as a result of gradual change involving elaboration of existing elements and the addition of new ones. Finally, such changes that do occur in a city, are reflections of both the needs of its inhabitants and the environment around it.

Now let us look at some animals.

An animal is like a city. At first glance it appears to be extremely complicated as can be seen in fig 4 (which is nevertheless grossly oversimplified). Closer inspection, however, reveals that an animal such as a fish is composed of a number of structural elements (fig 5). Like those of a city, each struc-

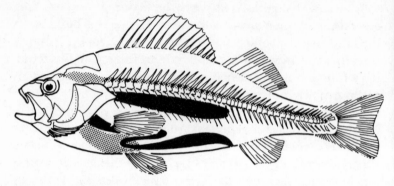

Fig 4 An animal, like a city, is a complex of structural (anatomical) and dynamic (physiological) elements

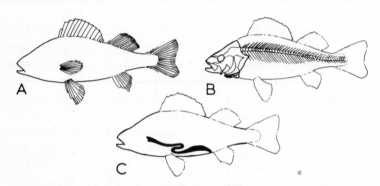

Fig 5 Some structural elements of a fish; fins A, skeleton B, and gut C. Each element has unique characteristics and functions

tural element has its own peculiarities and definite functions. Once again, like the elements of a city, the elements of an animal are integrated with one another so that none operates in isolation from the others. Indeed there are many close parallels between structural elements of cities and those of animals. One example is that of streets and blood vessels. The function of a street system is to provide a network along which people and things may travel or be carried from one point to

18

any other in the city. As a result, a street is a linear feature, and a street system a network. One-way systems ensure both that congestion does not occur, and that a continuous flow of traffic is maintained. The function of blood vessels is to provide a system enabling oxygen and other substances to be transported to every part of the body. They too are linear and structurally form a ramifying network. Because blood corpuscles are passive and must be pumped, blood vessels are almost always one-way; arteries lead from the heart, and veins lead to the heart.

In a similar way, physiological processes in the body may be likened to the dynamic elements of a city.

I do not want to push the above analogies too far but, before we can begin to understand evolutionary theory, it is essential that we visualise animals for what they are—complexes of closely integrated structural and dynamic elements.

At another level, we may liken the structure of a city to that of an animal community. Animals feed on plants and upon one another. In any community, there exists a structure of relationships called a food web. An example is given in fig 6 in which the animals and plants of the sea are represented as a system of elements that are dependent upon one another,

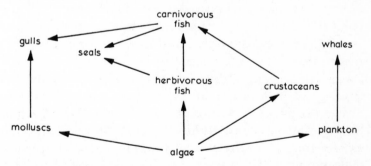

Fig 6 The feeding relationships in a community of animals and plants. As in a city, different elements are dependent upon one another for their livelihoods

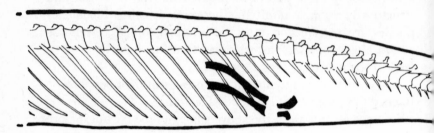

Fig 7 The skeleton of the python includes a pair of hind limbs that have lost their function. They are evidence that snakes evolved from animals that had limbs

and it can be seen that the result is similar to a system of dynamic relationships that might occur in any city. Relationships between members of a single species may be likened to the internal dynamics of a city, and those between different species to external dynamic relationships.

Just as it is difficult to conceive of the creation of a city *de novo*, it is impossible to imagine the spontaneous creation of animals and plants. The complex of elements that constitutes an animal is the result of millions of years of gradual change. Evidence for this assumption will be discussed later in this book, but, for now, a single piece of evidence may be mentioned. The python, like all snakes, appears to be legless, but in fact dissection shows that there are remnants of the hind limb skeleton enclosed within the body (fig 7). Snakes have evolved from lizard-like animals which had legs. In the course of evolutionary change these were lost, but, as we have seen, not entirely. The limb bones in the python are like Roman ruins in a modern town; evidence of a past history.

2

Themes

There are over seven hundred thousand different insects living in the world today, and every year several thousand new forms are discovered. Yet every insect is built on the same body plan. In every insect there is a head, a thorax, and an abdomen (fig 8). These divisions are easy to see because the body is enclosed within a layer or cuticle that supports the body and which is therefore called an exoskeleton. The thorax is always divided into three parts or segments, while the abdomen has as many as eleven.

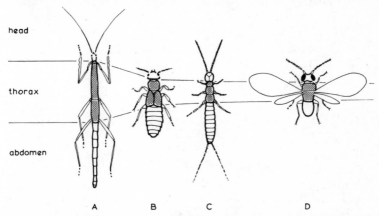

head

thorax

abdomen

A   B   C   D

Fig 8 All insects are built on the same basic plan. In all, the body is divided into three parts, the head, thorax, and abdomen. A, stick insect *Caransins*; B, beetle larva *Ocypus*; C, a dipluran *Campodea*; D, warble fly *Hypoderma*. Not drawn to scale

Vertebrates, the group that includes fish, amphibians, reptiles, birds, and mammals, have an endoskeleton that is enclosed within the body. Once again, no matter which vertebrate we choose to study, we find a common plan. The brain and sense organs are contained in a skull which articulates with a strong flexible backbone to which two pairs of fins or limbs are joined at the shoulder and hip girdles (fig 9).

If we look at the vertebrate skeleton in more detail we find more evidence of a common theme. Take the fore limb for example. Whether it be used for flying, swimming, walking, or grasping, it is composed of three long bones, a number of wrist bones, and as many as five jointed fingers (fig 10). The

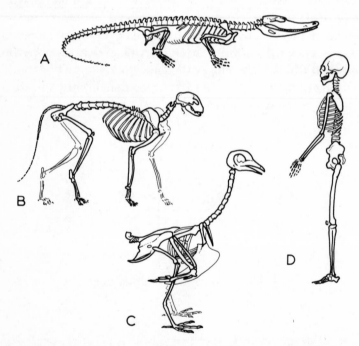

Fig 9 The skeletons of vertebrates are built on a common plan although modified for a variety of functions. A, crocodile; B, cat; C, pigeon; D, man. Not drawn to scale

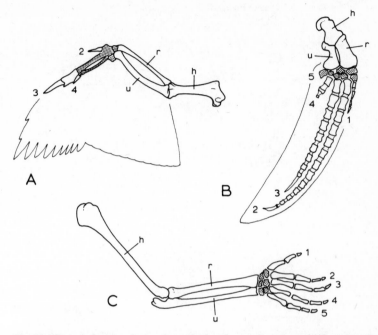

Fig 10 The skeleton of the fore limb of vertebrates modified for flying, swimming, and grasping. In each, the forearm is composed of three bones, the humerus h, radius r, and ulna u. The wrist bones in each are stippled, and the fingers numbered 1–5. A, pigeon; B, dolphin; C, man. Not drawn to scale

corresponding parts of the fore limb in different vertebrates are said to be homologous with one another. To recognise homologous parts in a group of animals is to acknowledge that its members are built on a common plan.

What of fossil forms? Fossils are the remains of animals (and plants) preserved in sands, gravels, and muds which have been deposited in lakes, rivers, and the sea during the earth's geological history. These deposits turn into rocks in course of time. Preservation of an entire animal is extremely rare, and usually only the hard skeleton is ever found. Let us compare two fossils with living animals. *Toxodon* (fig 11A) is a com-

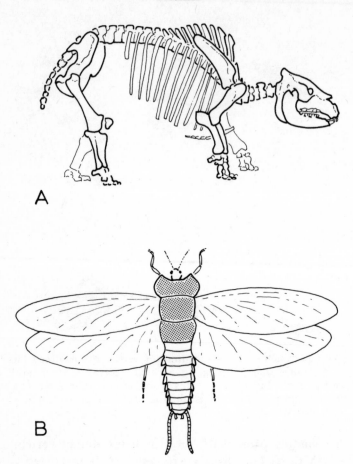

Fig 11 The basic body plans seen in groups of living animals can be identified in extinct forms. A, *Toxodon* is a vertebrate and B, *Homoioptera* an insect. Not drawn to scale

mon fossil from South American sedimentary rocks which are about 5 million years old. It is immediately apparent that *Toxodon* exhibits the same general characteristics as are rocks of the Carboniferous period in France which are over found in living vertebrates. Our second example (fig 11B) is called *Homoioptera*. It is like a dragonfly and is found in

350 million years old. Its body is divided into three parts, and it is obviously an early example of the theme exemplified by modern insects.

Paleontology (the study of fossils), shows that extinct animals are variations of themes discovered by comparing animals living today.

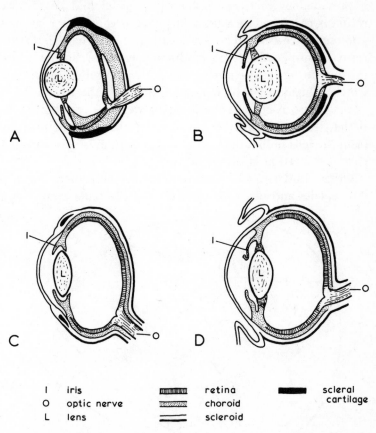

I	iris	▦	retina	▰	scleral
O	optic nerve	▨	choroid		cartilage
L	lens	═	scleroid		

Fig 12 Vertical sections through the eyes of four vertebrates show a remarkable uniformity of structure. In each case the eye ball is composed of three layers and the lens suspended from the choroid. A, salmon; B, frog; C, pigeon; D, horse. From Romer, *The Vertebrate Body*, 1962. With permission of W. B. Saunders and Co

Themes are not only illustrated by comparison of skeletal parts. The vertebrate eye is like a camera. In animals as diverse as a salmon, a frog, a pigeon, and a horse, it consists of a sphere composed of three layers (fig 12). The choroid and scleroid form protective and nutritive layers, the latter being transparent in front of the lens. Sometimes additional strength is provided by cartilage or bone in the scleroid layer. The retina is composed of a layer of pigment cells and a layer of photosensitive cells. The lens, which focuses light onto the retina, is suspended from a modified part of the choroid, while the amount of light entering the eye is controlled by the iris which is made of cells derived from both the choroid and retina layers. A curious similarity between all vertebrate eyes is that the light-sensitive tips of the retina cells point away from the lens so that light, before it is seen, must pass through part of the retina layer.

Other similarities between animals are even more curious. All reptiles and birds for example, lay the same kind of egg

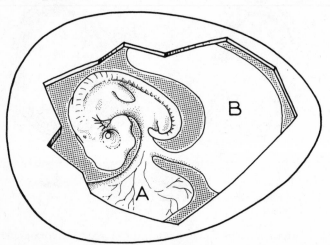

Fig 13 The eggs of crocodiles, turtles, and birds are built on a common plan. In each, the embryo derives food from a yolk sac A, and oxygen from the allantois B

(fig 13). In it, two sacs form as extensions from the embryo's gut. In one, the yolk sac, food for the developing crocodile, turtle, or bird is stored; the other, the allantois, acts both as a lung and a bladder. It extracts oxygen from the surrounding 'egg white' and acts as a receptacle for waste products.

Vertebrate embryos themselves are remarkably similar to one another. Compare those in fig 14, and remember how

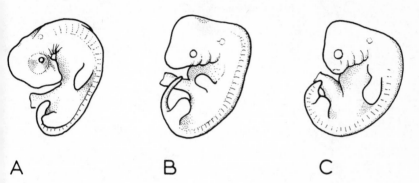

A   B   C

Fig 14 Even though the embryo of a bird A, is contained within an egg, it is remarkably similar to the embryo of a dog B, or of a man C. Not drawn to scale

different are the adult forms! We can sympathise with the nineteenth-century embryologist von Baer who, having preserved an embryo without labelling it, was unable later 'to remember whether it was a lizard, a bird, or a mammal.

A well-known parasite of crabs is called *Sacculina*. Its larva settles on its victim, pierces its body wall, and eventually becomes established underneath the crab's gut. From here, the adult *Sacculina* grows rootlets which ramify throughout the entire body of its host (fig 15A). The adult *Sacculina* is such a strange individual that it defies comparison with any other animal, but its larva (fig 15B) is very similar to those of crustaceans (the group that includes shrimps, crabs, and lobsters) (fig 15C). Note particularly that in both larvae there are

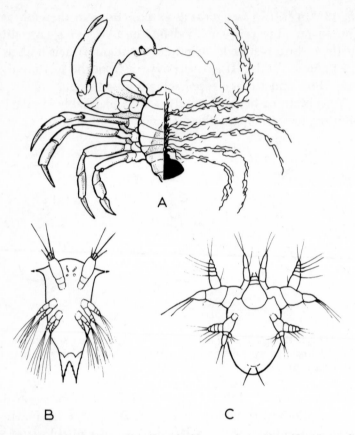

Fig 15 The parasite *Sacculina* lives inside a crab, and ramifies throughout its body, A. The relationship of *Sacculina* with crustaceans would
have been unsuspected if it were not for the similarity between its
larva, B, and that of crustaceans such as *Cyclops*, C. Not drawn to
scale

three pairs of jointed legs or appendages. We may conclude
that *Sacculina* is in fact a crustacean, a conclusion that would have
been impossible if we had been able to compare only adults.

However bizarre an animal may be, some stage in its life
history usually shows that it is no more than a variation of a
theme seen elsewhere.

Themes

Some similarities between animals are less obvious than those of gross morphology, but are just as significant. Our blood is red, and it is not too surprising to find that other vertebrates have red blood as well. The colour is due to the presence of a protein called haemoglobin which contains iron. The blood of crustaceans on the other hand, contains copper, and in consequence is blue.

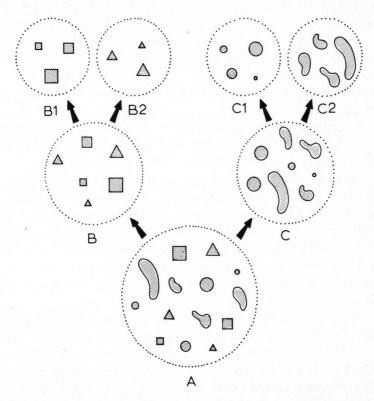

Fig 16 The classification of an assortment of shapes in group A, is made possible by the recognition of common characteristics. Group B for example, contains only shapes with straight edges, while further subdivision produces groups B1 and B2 which contain shapes with four and three sides respectively

And what do all these similarities between animals indicate? They indicate that animals do not differ from one another as do a thousand scraps of paper torn at random, but that animals are variations on a few basic plans or themes. Most importantly, the patterns seen in animal diversity suggest very strongly that evolution has taken place. If every animal had been individually created, one would not expect such patterns to emerge.

Because animals are variations on themes, classifications of animals are possible, but before discussing these, we must consider classifications of a more general kind.

Look at the collection of shapes in group A shown at the bottom of fig 16. At first they appear to be a random assortment but, by noting certain similarities between some of the shapes, they can be separated into two groups, B and C. B contains the shapes with straight edges and C those with rounded edges. Analysis of groups B and C shows that further subdivision is possible. B can be divided into group B1 (squares) and B2 (triangles); C can be divided into group C1 (circles) and C2 (irregular shapes). Two general comments can be made about this classification: firstly, the groups bring together shapes with similar characters, for example, group B1 contains shapes with four straight edges; and secondly, the members of each group can be regarded as variations of a single theme, for example, the squares in group B1 vary only in size.

If we now return to the animal kingdom, we find that animals can be classified in exactly the same way as could our shapes. Insects, for example, share certain general characteristics as do a collection of shapes with straight edges, and so they can be classified together. And just as a group of straight-edged shapes can be further divided into groups containing squares and triangles, so too can insects be subdivided, for example into one group in which wings are present, and another in which they are absent.

Another feature of classifications is that they contain hierarchies of groups. In our classification of shapes, it is clear that group B is different to groups B1 or B2 because it is larger and contains a greater variety of shapes; it has a higher position in the hierarchy of groups. Likewise, group B is below group A in the hierarchy because it contains a smaller variety of shapes than does group A. In other words, there are three

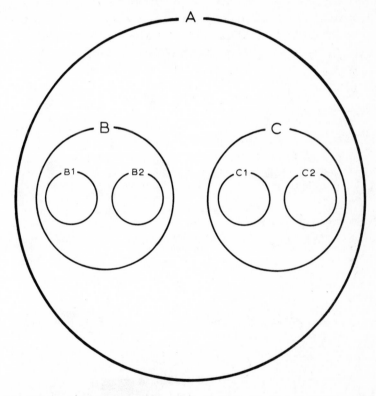

Fig 17 The hierarchy of groups defined in fig 16. Shapes in group B2 are automatically contained within groups B and A. Likewise, shapes in group C1 are contained in groups C and A. Group A is high in the hierarchy of groups, while groups B1, B2, C1, and C2 are the lowest members of the hierarchy

kinds of groups in our example: a large group with general characteristics, group A; moderately sized groups, groups B and C; and small groups with well-defined characteristics, groups B1, B2, C1, and C2. The hierarchy of groups in this classification is thus a simple one, involving three types of group and two levels at which they are subdivided. The relationship between the groups is shown in fig 17, which illustrates how, for example, group B contains groups B1 and B2, while being contained within group A.

An example from zoology: a group called the Anthropoidea contains the monkeys, apes, and man; the Anthropoidea, however, is but one of many groups contained within the class

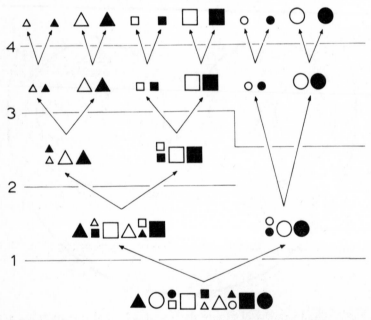

Fig 18 The classification of a group containing twelve different shapes into twelve groups each containing a single shape involves subdivision of groups at four levels. At each level, the groups become more and more closely defined

Page 33 (right) Fossilised remains of a trilobite, *Ogygiocarella debuchi*. In spite of the fact that this animal died almost 500 million years ago, its structure is based on the same general body plan seen in living crustaceans such as the woodlouse; *(below)* When a fish is fossilised, only the skeleton, and sometimes the scales and teeth, are preserved. Comparison with living fish shows that this specimen, *Eohutus minimius*, had characters that are very similar to those of fish alive today

Page 34 The forelimbs of birds and bats have been modified to make flight possible. In birds, the flight surfaces are composed of feathers, while in bats, the wings consist of a skin stretched between the fingers. These and other differences show that birds and bats are members of separate groups although they are variations of the same general body plan typical of the vertebrates

Mammalia, which includes anthropoids, carnivores, rodents, elephants, bats, and many more.

A more varied selection of shapes is classified in fig 18. A single large group has been divided into twelve small groups as a result of division, or classification, at four successive levels. At the first level, straight-edged shapes are separated from rounded shapes; at the second level, the straight-edged shapes are divided into groups containing three- and four-sided shapes. As the rounded shapes do not differ in the number of sides they possess, they cannot be divided at this level. At the third level, small shapes are separated from large shapes. Finally at the fourth level, the colour difference, black and white in each group, is used as a criterion for a separation into small groups with definite characteristics with regard to shape, size, and colour.

It should be admitted that the classification outlined above is not the only one possible. For example, black shapes could have been separated from white shapes at the first level of classification, and considerations of shape and size left till later. However, if this were done, one would still eventually finish up with the same twelve small groups.

We have seen that classification involves successive subdivision of groups so that a large group with general characteristics is ultimately classified into a number of much smaller groups with very definite characteristics. Exactly the same thing happens in animal classification. Figure 19 is a classification of animals with special reference to frogs. At the first level of classification animals with three characters in common are separated from all other animals. These characters are a skeletal rod called a notochord which runs along the back of the body, a similarly placed nerve cord, and gill slits; the group into which all animals with these characters are placed is called the Chordata. At the second level, the animals with a well-developed brain and an endoskeleton are separated from other chordates. This group contains the vertebrates, the

35

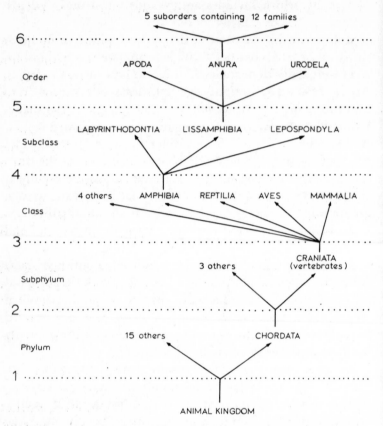

Fig 19 The classification of frogs. At each level in the classification, groups are defined with increasing precision. At level 5, the frogs (Anura) are separated from all other animals. Continuation of such a classification would end with the definition of every species of frog

group discussed earlier in this chapter. At the third level, the vertebrates are divided into eight groups, four of which are sometimes lumped together and called fish, while the others contain the amphibians, reptiles, birds (Aves), and mammals. The amphibians are vertebrates which typically spend part of their lives in water and part on land. They are divided into

Themes

three groups at the fourth level of classification. Two of these three groups, the Labyrinthodontia and the Lepospondyla, are only known as fossils. At the fifth level, the living amphibians, the Lissamphibia, are divided into three groups: the Apoda which are legless burrowing forms; the Urodela (or Caudata) which have tails, and which include the salamanders and newts; and the Anura which lack tails, and which include the frogs and toads.

What have we done by classifying frogs? Well, we have outlined a hierarchy of groups which starts with large ones with general characteristics (the animal kingdom), and finishes with small, well defined groups (two species of frog for example, may differ only in the pattern of their skin markings). More importantly, we have identified a number of themes. Frogs are variations of the vertebrate theme, that is, frogs are distantly related to all other vertebrates. As we pass down the hierarchy of groups we define the relationships of frogs more and more exactly. The fact that frogs are variations of the Amphibian theme means that frogs are more closely related to other amphibia than they are to other vertebrates.

In zoology, each type of group in the hierarchy produced by classification is given a name. The largest groups with general characteristics are called phyla (singular—phylum), and represent the basic themes seen in the animal kingdom. There are only sixteen such groups. Variations of phyla are called subphyla, variations of subphyla are called classes, variations of classes are called subclasses and so on. At the other end of the hierarchy are the smallest groups, these are called subspecies. The main group types used in zoology are, starting with the largest:

> Kingdom
> Phylum
> Subphylum
> Class
> Subclass

Order
Suborder
Family
Subfamily
Genus
Species
Subspecies

The full name of any animal would include the name of every group, or taxon, to which that animal belongs. Hence the common frog found in Great Britain is classified and named as follows:

Kingdom	Animalia
Phylum	Chordata
Subphylum	Craniata
Class	Amphibia
Subclass	Lissamphibia
Order	Anura
Suborder	Diplasocoela
Family	Ranidae
Genus	*Rana*
Species	*temporaria*

In practice, however, only the genus and species names are used, and the common frog is called *Rana temporaria*. This simply means that of the theme recognised as the genus *Rana*, the common frog is one variation called *temporaria*.

I have tried to show that variation of animal structure is not as random as at first appears. Large groups of animals, both living and fossil, contain variations of a limited number of themes, and this makes classification of the animal kingdom possible. Classifications are composed of hierarchies of groups. Any one group represents a theme, and the groups which result from division of that group are variations of that theme.

3

Distribution in Space and Time

In the last chapter we saw that there are patterns underlying animal variation. Here we will discuss the relationships between these patterns and the geographical distribution of animals, and their occurrence during the earth's geological history.

Everyone knows that different animals and plants occur in different parts of the world. A few forms, and man is a good example, are ubiquitous, others are limited to a single continent, to a region such as a mountain range or desert, or to a single island or lake. The geographical distribution of animals is an important clue about the evolutionary process, and I want to discuss two examples; one involves the distribution of a large taxon over a large area, while the other involves the distribution of a small taxon on a small group of islands.

The class Mammalia is the group of animals characterised by, among other things, a large brain, the presence of fur, and an ability both to control the temperature of their blood (they are sometimes misleadingly called warm blooded for this reason) and to secrete milk for the nourishment of their young. The mammals include most well-known animals including cats, dogs, horses, cows, elephants, monkeys, and man. There are three variations on the mammalian theme which enable the mammals to be classified into three groups, the mono-

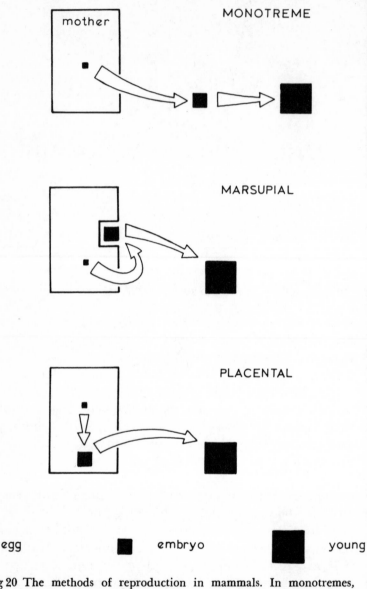

Fig 20 The methods of reproduction in mammals. In monotremes, only the egg is retained inside the mother. Marsupials retain the embryo in a pouch, and in placentals, the young are born at an advanced stage of their development

tremes, marsupials, and placentals. The monotremes and
marsupials are each included within a single order while the
placentals are more varied and can be divided into twenty-five
orders. Each group has a distinctly different method of reproduction (fig 20). The monotremes, which include the duck-
billed platypus and the spiny anteater, lay eggs, so that the
embryo develops outside the mother. The eggs of marsupials
on the other hand are retained inside the mother for a short
period, only eight days in the case of the opossum, and the
young are born in a very immature or embryonic state. The
embryo immediately climbs up its mother's belly to a pouch
where it completes its development. Finally, in the placental
mammals, the embryo is retained within the mother for a long
period.

When it is born, an animal is especially vulnerable because
it is so helpless. Figure 20 shows that the three mammalian
methods of reproduction are progressively more efficient because there is a trend towards retention of the early stages
of the offspring's life history in a place where there is
most protection from the hard outside world, inside the
mother.

The marsupials include the kangaroo, the banded anteater,
the tasmanian wolf, the opossum, and the koala bear. The
remarkable fact about most of these animals is that in their
appearance and way of life, they are closely paralleled by
mammals that are variations on the placental theme.

MARSUPIAL	PLACENTAL
tasmanian wolf	wolf, dog, fox
banded anteater	anteater
marsupial mole	mole
pouched mouse	mouse
flying opossum	flying squirrel
marsupial jerboa	jerboa
tasmanian tiger cat	cat
rat kangaroo	rat

41

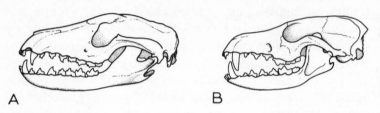

A   B

Fig 21 The skulls of the tasmanian wolf, A, and the fox, B. Each animal belongs to different groups of mammals (marsupials and placentals), but are remarkably similar to one another because they have similar feeding habits. Not drawn to scale

The close similarity between members of each of these pairs and a fox are compared in fig 21. Both animals are flesh eaters, is impressive. For example, the skulls of the tasmanian wolf and in both the snout is long, while the jaws are powerful and equipped with back teeth modified for shearing and front teeth for tearing and for piercing. Fig 22 is a drawing of a

Fig 22 An Egyptian wheelwright from a painting at Thebes, about 1500BC. It comes as no surprise that the shape of the wheel is the same as those used today. At all times, function determines form. Based on a drawing from Derry & Williams, *A Short History of Technology*, 1960

wheelwright taken from a tomb painting at Thebes which is almost 3,500 years old. It comes as no surprise that the wheel in the drawing is circular for, whether made in 1500 BC or today, a wheel performs a function which is only served if it is circular. In the same way, it is not too surprising that the tasmanian wolf and fox are similar, for the skull in each performs similar functions. If it were not known that the tasmanian wolf had the marsupial method of reproduction it would certainly be classified with the placental dogs, wolves, and foxes. The marsupials then, are similar to placentals, yet have what may be regarded as a less efficient method of reproduction. When the internal combustion engine was invented, horse-drawn carriages soon became extinct. So why haven't the marsupials become extinct in the face of more efficient competitors?

The answer is that marsupials are protected by their isolated distribution, for, although their fossils have been found in North and South America, Europe, and Australia, living forms

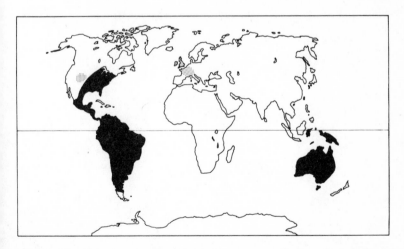

Fig 23 Closely related animals tend to have similar geographical distributions. The distribution of living marsupials (black) and fossil forms (stipple)

only occur in South America and Australia, and most of them in the latter continent (fig 23). How marsupials got to Australia will be discussed in the next chapter.

Our second example concerns a group of small birds which, because they all have the same grey or brown colour, short tails, similar courtship displays, nests, and eggs, are classified together in the same subfamily, the Geospizinae. The Geospizinae are found only on the Galapagos Islands and the island of Cocos (fig 24). The Geospizinae, or Darwin's finches as they are commonly called, are not only similar to one another but also to forms which live on the South American mainland. They are classified into five genera which contain a total of fourteen species (fig 25). These species differ from one another in their feeding habits, and these differences are reflected in the size and shape of their beaks. The genus *Geospiza* contains six species, the ground finches, that live on the ground and feed mainly on seeds. Forms such as *G. difficilis* feed on small seeds and have small beaks, while *G. magnirostria* feeds on larger and harder seeds and has, not unexpectedly, a larger, stronger beak. The second genus,

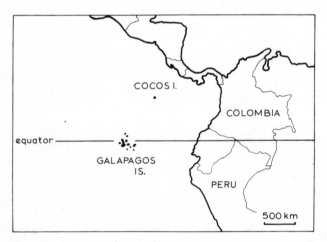

Fig 24 The isolated location of the Galapagos and Cocos islands

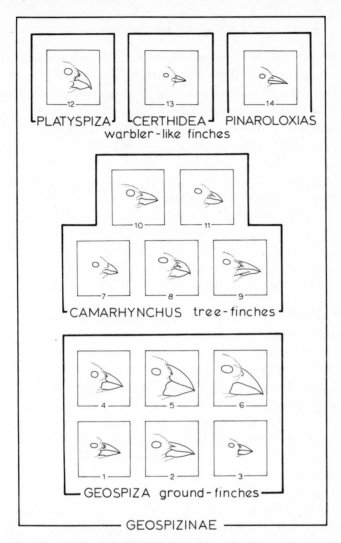

Fig 25 The classification of Darwin's' finches. All are contained within the family Geospizinae which contains five genera. The variety of form of the bills in these birds is related to their feeding habits. 1 *G. difficilis* (seeds), 2 *G. scandens* (seeds, flowers, and fruit), 3 *G. fuliginosa* (seeds and some insects), 4 *G. fortis* (various seeds), 5 *G. magnirostris* (large seeds), 6 *G. conirostris* (cactus), 7 *C. parvulus* (small insects), 8 *C. psittacula* (large insects), 9 *C. pallidus* (large insects), 10 *C. heliobates* (large insects), 11 *C. pauper* (medium sized insects), 12 *P. crassirostris* (buds, leaves, and fruit), 13 *C. olivacea* (small insects), 14 *P. inornata* (various foods). All drawn half natural size. Based on a drawing from Lack, *Darwin's Finches*, 1947

Camarhynchus, has five species which live in trees and feed on insects. Among them is the famous woodpecker finch *C. pallidus,* that uses a twig as a tool with which to probe for insects. The three other genera each contain a single species. *P. crassirostris* feeds on buds, leaves, and fruit, and has a heavy beak with which it crushes its food, *C. olivacea* has a small delicate beak for picking insects off leaves, while *P. inornata* has a pointed beak, and feeds on a variety of food including insects, and is found only on the island of Cocos.

The geographical distribution of Darwin's finches, like that of the marsupials, poses problems. In this case the obvious questions are; why are there so many different varieties based on a single limited theme, and why are these variations found only on a small group of isolated islands in the eastern Pacific Ocean?

The next two examples involve the distribution of groups of animals, not in particular areas, but over particular periods of time.

The taeniodonts are fossil mammals which have been found in rocks of Palaeocene and Eocene age in North America. Five genera are similar to one another and are classified in the same subfamily called the Stylinodontinae. Unfortunately the four well-preserved genera do not have common names and I must refer to them by the scientific names which are: *Wortmania, Psittacotherium, Ectoganus,* and *Stylinodon.* Their skulls are shown in fig 26. It can be seen that the differences between the four genera are of a progressive kind. The skulls become larger, and the lower jaw deeper and more robust. The saggital crest, which is a ridge of bone running along the top of the back of the head, becomes higher. Relative to the rest of the skull, the snout becomes shorter. Other differences are seen in the teeth. *Wortmania* has fairly small teeth although the canines, the pointed teeth at the front of the mouth, are well developed. At the other extreme, *Stylinodon* has enormous canine teeth which are especially interesting because they

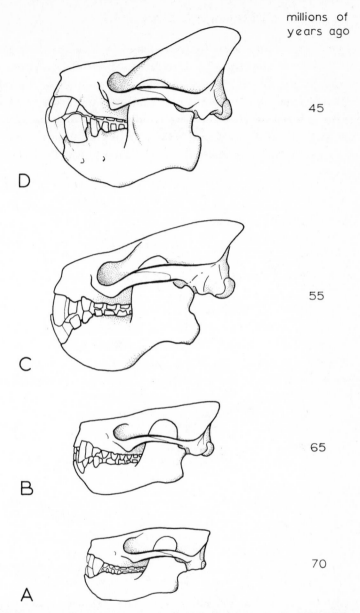

millions of
years ago

45

55

65

70

Fig 26 Variation of skull form in the taeniodonts. During 25 million years, the taeniodont skull became larger and the teeth more robust. A *Wortmania*, B *Psittacotherium*, C *Ectoganus*, D *Stylinodon*. Drawn to scale. Based on a drawing from Patterson, in Jepson, Simpson, and Mayr, *Genetics, Paleontology and Evolution*, 1963

have enamel only on their front edges. Moreover, their lack of roots shows that these teeth were not replaced as happens in many animals, but grew continuously as do those of rats, squirrels, and other rodents. Many animals with continuously growing teeth feed by gnawing. The rate at which the top of the tooth is worn down equals the rate of growth of the tooth.

As can be seen in fig 26, the four taeniodont genera can be

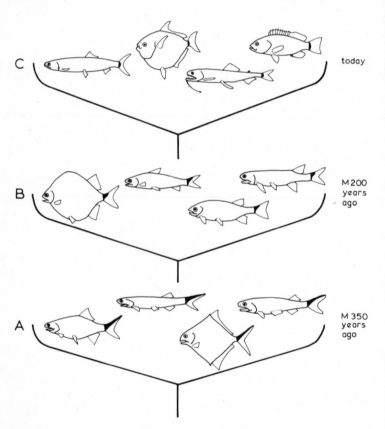

Fig 27 During the last 400 million years, bony fish have been characterised by three distinct types of tail: the heterocercal tail A, the hemi-heterocercal tail B, and the homocercal tail C. The homocercal tail is found in almost all bony fish living today

arranged in a definite series in which the features mentioned above are progressively developed. It is highly significant that the order in which our four genera occur in this series is exactly the same as the order of their occurrence in the earth geological record: *Wortmania* and *Psittacotherium* occur respectively in the Lower and Middle Paleocene, while *Ectoganus* occurs in the Lower Eocene and *Stylinodon* in the Middle Eocene. In all, the series spans a period of at least 25 million years.

An example of variation over a greater period of time, 400 million years, is provided by the bony fish. During the Mesozoic period, which occurred between 250 and 400 million years ago, there was a great variety of fish. Their body shapes varied from slender to deep, but in all there was a heterocercal tail, that is, a fleshy lobe extended along the entire length of the upper part of the tail (fig 27). Today, with one or two exceptions, this lobe is never present although there is quite as much variation in body form. An intermediate condition, in which the fleshy lobe is only partly developed, is typical of fish that lived during the Triassic period about 200 million years ago.

A simple analogy helps explain what has happened to fish during the past 400 million years (fig 28). Two hundred years ago, ships were typically dependent upon the wind whatever their shape and function. Today, steam and the internal combustion engine are almost universally used. During the middle part of the last century, many vessels were built that were intermediate between the extremes of sail and steam, and ships such as the *Great Eastern* used both methods of locomotion. The suggestion thus occurs to us, that the tails of fish were gradually being modified, and that the change in their structure is due to the perfection of an increasingly efficient type.

We have seen that animal variation is not haphazard but represents variation of a number of themes. We now see that

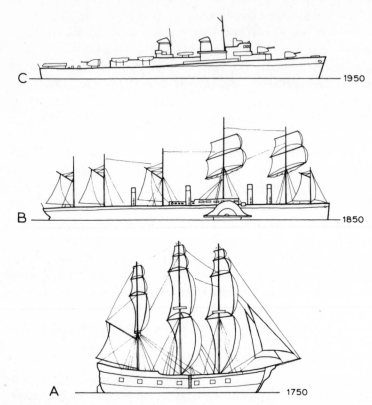

Fig 28 Just as hemiheterocercal tails are intermediate in form between those of ancient and modern fish, ships such as the *Great Eastern*, B, are intermediate in construction between sailing ships, A, and screw-driven ships of today, C

some themes and their variations are limited to definite geographical regions, while others occur over definite periods of time. Moreover, these observations can be refined, for, as we know, animal themes can be equated with taxons or groups, used in classifications. In general, the higher a taxon is in the hierarchy, the wider is its geographical range and the longer is its period of existence. Hence, members of the class Mam-

Page 51 The marsupial mouse (above) and the Indian spiny mouse (below) are superficially extremely similar. Their methods of reproduction are, however, quite different—evidence that these animals belong to groups that have evolved separately for almost 100 million years

Page 52 The origin of great sea-monsters and every kind of bird as depicted in the Cambridge Bible of 1663

malia are found all over the world, but the genus *Phascolarctos*, the Koala bear, is found only in parts of Australia. The subfamily of taeniodonts called the Stylinodontinae have a known time range of about 25 million years, but the genus *Wortmania* occurs only in the Lower Paleocene, a period of about 5 million years.

4

The Idea of Change

Animal diversity, geographical distribution, and occurrence during periods of the earth's history, are all related to a single phenomenon, change. Unless animals are assumed to have changed, nothing discussed in the last two chapters makes sense, but if we acknowledge change to have taken place, a single unifying theme begins to emerge.

Let us first consider simple change. In fig 29 a series of squares has been drawn. Two changes are apparent as we pass from the bottom to the top of this series, the squares become larger, and the black triangular area increases in size. One complicating factor in this series is that the rate at which the black area increases in size is faster than the rate at which the squares become larger. Thus, although at the beginning of the series only a tiny corner of the square is black, by the end of the series half the square is black. The change in this series is gradual.

If we consider a sample from the series, squares A–D, we may still reasonably infer that change has taken place, but there is less evidence that the change was gradual. If we know that squares A–D are in fact a sample from a longer series, it is not too difficult to imagine the missing elements and to reconstruct the complete series from which the sample has been taken.

Reconstruction of series from samples is what paleontologists spend much of their time doing. A good example of a

series   sample

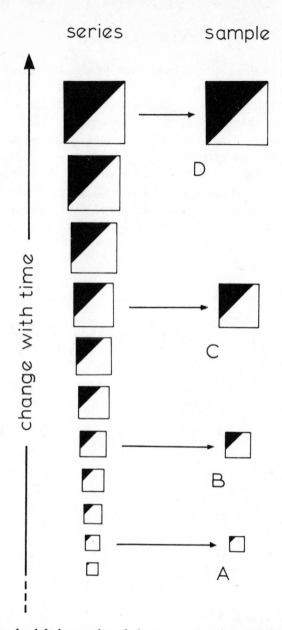

change with time

Fig 29 On the left is a series of shapes representing simple change. Two changes are taking place, both the squares and the black triangles are becoming larger. The rate of change of the triangles is faster than the rate at which the squares are becoming larger. A–D represent a sample from the series. Study of the sample would enable us to deduce that simple change had taken place

paleontological sample was seen in the last chapter. It is reasonable to assume that the four skulls in fig 26 (p47) are a sample from a series in which gradual change has taken place. Like the shapes in fig 29, different parts of the taeniodont skulls have changed at different rates. This can be seen by comparing the extreme members of the sample, *Wortmania* and *Stylinodon*. Change between these two members has resulted in an increase in overall skull size by a factor of 1·5, that is, *Stylinodon* has a skull one and a half times long as has *Wortmania*. But proportionately, the canine teeth of *Styli-*

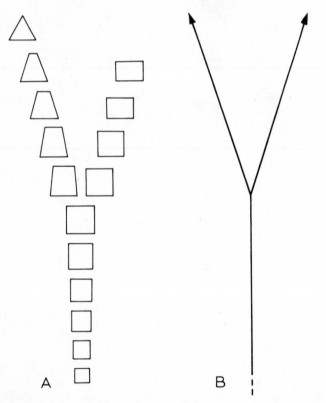

Fig 30 Divergent change in a series of shapes A is represented by the arrows in B

nodon are much larger than in *Wortmania*, in fact the canines have increased in length by a factor of at least 4.

Because the chances of fossils being preserved at every stage of the earth's history are slight, the paleontological record of animal change is never quite complete. The paleontologist always works with a sample of any series of animals. The example of the taeniodonts shows that there is sometimes enough material available for paleontologists safely to infer that gradual change has taken place. It should be stressed that, although individuals have been portrayed in fig 26, the changes we have just discussed are only properly understood in terms of whole populations of animals. This aspect of change will be discussed further in Chapter 6.

More complex than simple change is divergent change. In fig 30 we begin with a small square. As a result of simple change it becomes larger, but at a certain point divergence takes place. In this example, the sides of some squares become inclined, finally to produce a triangle, and in others they shorten to produce a rectangle. Divergence can be simply illustrated by the letter Y. Movement from the bottom to the top of the Y indicates change during a particular period of time, while the final separation of the limbs of the Y symbolises the divergence that has taken place. Whereas simple change can cause one animal to develop into another, divergent change must take place before one animal can produce two or more different forms.

Fossil evidence strongly suggests that divergence took place about 250 million years ago to produce two closely related families of fresh-water fish (fig 31). In one family, the Brookvaliidae, there is a bone called the premaxilla (stippled) at the tip of the snout, in another, the Redfieldiidae, this bone is missing. Apart from this difference, the two families are very similar to one another, especially because in both, a single nostril lies behind the nasal bone. In earlier fish, there are two nostrils, one of which lies in front of the nasal bone. It is

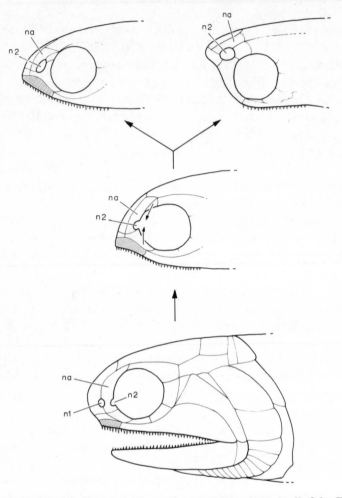

Fig 31 Simple and divergent change in a group of fossil fish. The families Brookvaliidae and Redfieldiidae have diverged from a common ancestor. In the former, a bone called the premaxilla (stippled) is retained, while in the latter it is lost. Prior to divergence, simple change has produced a form in which the front nostril (n1) is lost or has joined the posterior nostril (n2), and in which the nasal bone (na) is separated from the edge of the eye socket by bones growing from above and below. Based on a drawing by Hutchinson, 'A Revision of the Redfieldiiform and Perleidiform fishes from the Triassic of Bekker's Kraal (South Africa) and Brookvale, (New South Wales),' *Bulletin of the British Museum (Natural History), Geology*, Vol 22 No 3, 1973

likely that, as a result of gradual change, the front nostril was either lost or became associated with the posterior nostril to produce a fish with a single nostril lying behind the nasal bone. Changes in the shapes of the bones above and below the nasal took place at the same time, eventually leading to the separation of the nasal bone from the edge of the eye socket. Later, divergence took place producing one group in which the premaxilla was gradually lost, and another in which it was retained.

In both the example of simple change, the taeniodonts, and divergent change, the brookvaliid and redfieldiid fishes, the changes that have been discussed can be related to feeding habits. The taeniodonts gnawed their food and so increase in size of the canine teeth was an obvious advantage. In the case of the fish, the loss of the premaxilla in redfieldiids is associated with sucking up food from muddy bottoms of lakes and rivers, while the brookvaliids had more varied feeding habits.

At this point it is necessary to introduce the words ancestor

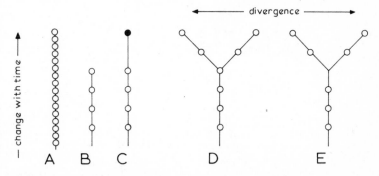

Fig 32 Graphic representations of simple and divergent change. A shows a series of fossil animals in which simple change is taking place over a period of time, B a sample from such a series, C a sample that includes a living form (black circle). Divergence is depicted by lines that become increasingly separated during the course of time. D represents divergence from a known fossil ancestor, and E divergence from an unknown ancestor

and descendant. In a series in which change takes place, the earliest known form is said to be ancestral to later forms. Later forms are said to be descended from earlier ones. In the case of divergent change, two or more descendants have common ancestry because they can be traced back to a single common ancestor.

Various combinations of simple and divergent change can produce an almost endless variety of patterns. The simplest ones are shown in fig 32. Simple change in a series of fossils can be represented by a row of open circles A. But, as we have seen, the fossil record usually provides only a sample from such a series, and this situation is shown in B, in which each circle represents a fossil, and the straight line, inferred gradual change. Thus B represents the situation seen in the taeniodonts (fig 26). If there was a taeniodont still alive today, and if it was obviously descended from the known fossil forms, we could represent the situation as in C, in which the living form is represented by a black circle.

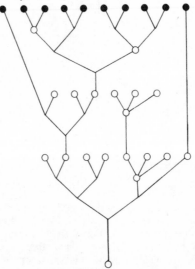

Fig 33 A family tree of living and fossil forms, that have originated by simple and divergent change from a single ancestor

Divergent change is also graphically represented in fig 32. In D we have the situation where the common ancestor to two diverging lines is known. More commonly, the actual ancestor is not known and we can only infer its presence, E.

More complex patterns than those just mentioned do occur, but in fact, they always contain the same elements of simple and divergent change. Such patterns are sometimes called 'family trees'. The pattern or 'family tree' in fig 33 is complex and includes both fossil and living forms. It can be seen that all the individual types represented can be traced back to a single ancestor; for this reason the diagram represents what is known as a radiation of types. Examples of radiations have already been noted, they were the marsupials and Darwin's finches.

Figure 33 is worth detailed analysis for it demonstrates how change and classification are related to one another. All the types represented in fig 33 are related to one another because, as I have said, they are all descended from a single ancestral type. They can all be considered, therefore, as variations on a single theme and so can be classified into a single large taxon. Let us suppose that the taxon in this case is a family, and let us call it A. As a result of divergence, family A contains two distinct groups, these would be subfamilies, B and C say, and they have been indicated in fig 34. All the forms contained within subfamily B can be considered to be variations of one theme, whereas those of C are variations of another. Further subdivision into genera is now possible, B into B1, B2, B3, and B4, and C into C1, C2, and C3. In each case the genus is defined by a group of forms that have a common ancestor and so have characters in common with one another. In fig 34, this is demonstrated by the fact that only one line ever enters an area defined as a genus. We could not include the living form in genus B2 within B3, because it is clear that it cannot be traced back to the common ancestor of all the forms in B3.

The circles in fig 34 represent species, and it can be seen

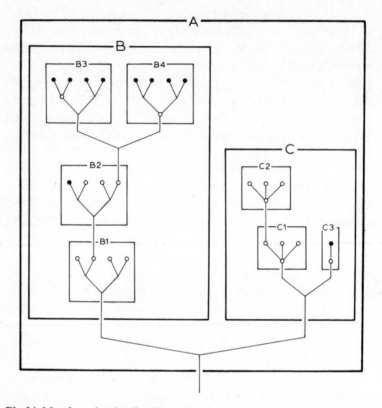

Fig 34 Members in the family tree seen in fig 33 can be classified as indicated here. All members are contained within a single family A which is subdivided into two subfamilies B and C. Subfamily B contains genera B1–B4 while subfamily C contains genera CI–C3. For explanation, see text

that some genera (B2, B3, B4, and C3) contain living and fossil species, while others (B1, C1, and C2) contain only fossil forms. Genera can, and often do, contain only living forms, but such an example does not occur in fig 34.

Often, taxonomists (people who classify animals and plants) are only interested in the relationships between larger groups. Figure 34 can be simplified and this has been done in fig 35 in which genera are regarded as units in the pattern. It should be stressed that in this case a line joining two genera simply indicates relationship between those genera and not change of any particular species.

So far, I have shown how change accounts for diversity in animal types, and I have also tried to show that the way in which a group of animals is classified reflects the way in which we think they are related to one another. If the family tree in figs 33, 34, and 35 seems rather academic, let me stress that all I wish to show can be expressed in very few words. Animals are altered over long periods of time by simple and divergent change. As a result, animals can be said to be related to one another. If two animals are closely related, a good classification places them in a taxon low down in the hierarchy; if they are only distantly related they are classified in different taxa

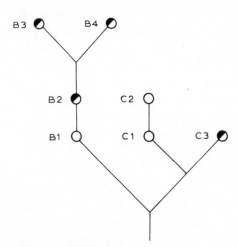

Fig 35 A simplified version of fig 34 in which only the relationships between genera are indicated

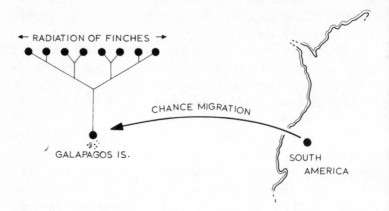

Fig 36 The various finches found on the Galapagos Islands are the result of divergence from a single species, a chance migrant from the mainland of South America

initially, but will end up in the same taxon higher in the hierarchy.

What of geographical distribution? In the last chapter we saw that some taxa are limited to a single group of islands or to one continent, also that taxa higher in the hierarchy tend to be widely distributed, while lower taxa are usually more limited in their distribution.

To begin with Darwin's finches. It will be remembered that all these birds, although differing from one another in their feeding habits, are similar in other respects and are classified into a single subfamily. We can now guess that Darwin's finches therefore represent the results of a radiation from a single ancestral type. Presumably at some stage a flock of finches from South America somehow made the 500 mile journey to the Galapagos. Once there, divergent change of this immigrant population produced a variety of birds that could take advantage of the different foods available on the islands (fig 36). The initial immigration was almost certainly a chance

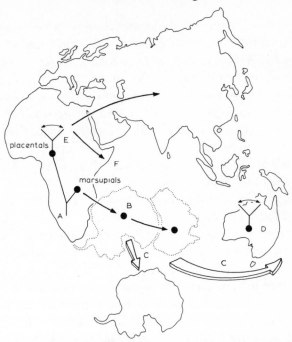

Fig 37 Why there are marsupials on Australia but no placentals. Marsupial and placental mammals originated from a common ancestor A, the former group diverging first. Because Antarctica and Australia were then joined to Africa, the marsupials could migrate overland to Australia, B. Later, after the continents had drifted apart, C, the marsupials underwent an adaptive radiation D. Meanwhile the placentals also became diverse, E, but could not migrate in the direction of Australia beyond the coasts of Africa and Asia, F

event, for very few other birds have found their way to the Galapagos.

And what of the curious distribution of the marsupials mentioned in the last chapter? The events leading to their isolation on Australia are summarised in fig 37. In Africa about 100 million years ago, the early mammals A diverged into at least two groups, among them the placentals and marsupials. We must assume that the marsupials diverged somewhat earlier than did the placentals. At that time, the

Antarctic and Australian continents were joined to Africa (dotted outlines), so that the marsupials could easily migrate to Australia B. Later, Antarctica and Australia drifted to their present position C, leaving the marsupials to produce by divergent change the variety we know today D. Meanwhile, the placentals also diverged and spread throughout Africa and Asia E but, because by that time Australia had drifted westwards, they were unable to reach that continent F. Some placentals have, in fact, since found their way to Australia. Forms such as dogs, rabbits, cattle, and sheep are recent arrivals, having been introduced by man. The only others are rats and bats. The rats all belong to one family, the Muridae, and they probably got to Australia by passing along the chain of islands that joins the Australian region to Asia. One can imagine the occasional chance event by which a few small animals are ferried across the water gaps on rafts of driftwood and vegetation. Bats present less of a problem because, of course, they can fly. There would be no difficulty for them to spread gradually from Asia along the Borneo, New Guinea island chain.

In conclusion, the idea of change helps us to understand many otherwise curious facts about animals. It explains why animals are similar to one another and why classifications can be made in which groups reflect relationships. It explains why series of similar animals occur in different horizons of the earth's sedimentary rocks, and why groups of similar animals are often found in certain parts of the world today. In fact, if we do not accept that animals have changed, and are changing still, we cannot begin to appreciate the nature of life on this planet.

5

The Emergence of
Evolutionary Thought

The idea of change took many years to become established in
men's minds. In 1484, Leonardo da Vinci (1452–1519) super-
vised the building of canals in northern Italy. While excavations were in progress, he noticed that remains of shells, crabs,
and fish were revealed. These were obviously marine creatures,
yet the excavations took place many miles from the sea. With
amazing perception Leonardo made the following deductions:
once, northern Italy had been submerged by a sea; in that sea
there lived mussels, oysters, scallops, crabs, and fish, just as in
modern seas, and over a period of many years, their remains
sank to the sea bed where they were entombed in sand and
mud. When the sea receded, these fossils remained, preserved
in what had become the rock forming the plains of Lombardy
(fig 38 A–C). Leonardo also explained how fossils were turned
to stone, a magical event in most people's eyes at the time. He
supposed that shells and other remains were first entombed in
sand or mud (fig 38D). When this sediment solidified, and the
organic remains had dissolved away, the only evidence of past
life were natural moulds in the rock (fig 38E). Later sediments
filtered through the rock and filled these moulds, producing
perfect replicas of the animal remains (fig 38F).

Unfortunately, Leonardo's speculations were not published,
and we only know of them today because his notebooks have

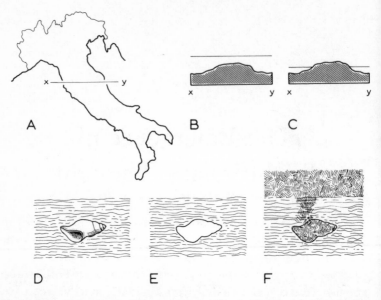

Fig 38 The deductions of Leonardo da Vinci. Because marine fossils were found in northern Italy, A, Leonardo realised that the land had once been submerged under a sea, B, that had later receded, C. Shells and other animal remains were entombed in sand and mud on the sea bed, D, and then rotted away leaving natural moulds, E. These were filled by later sediments, F, so that replicas of past life were formed in stone

been preserved. Two hundred years later a Dane called Nicolous Steno (1638–86), who also worked in northern Italy, made a great discovery. Common fossils around Florence were glossopetrae or 'tongue stones' (fig 39). Nobody knew what they were until one day in 1667, a shark happened to be stranded on the Italian coast. It was sent to Steno who compared specimens of 'tongue stones' with the shark and realised that the fossils were in fact shark's teeth. Moreover, he noticed that, although very similar to modern shark's teeth, the 'tongue stones' displayed minor differences. Steno concluded that 'tongue stones' represented the remains of a shark that was no longer in existence.

68

The next step towards an understanding of past life was made in 1746 by a Frenchman, Jean-Etienne Guettard. He studied the sedimentary rocks in the Paris region and saw that they were arranged in layers. For reasons summarised in fig 40,

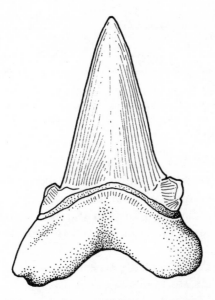

Fig 39 A source of bewilderment for early natural historians were 'tongue stones' or glossopetrae. Steno was the first to realise that they were fossil shark's teeth. This specimen is called *Lamna obliqua* and is about 50 million years old. Drawn at natural size. From *British Caenozoic Fossils*, British Museum (Natural History), 1963. With permission of the Trustees of the British Museum (Natural History)

he realised that the lower layers were older than those nearer the surface. He also noted that successive layers of rock contained different fossils and deduced that there had been gradual replacement of faunas during the history of the earth.

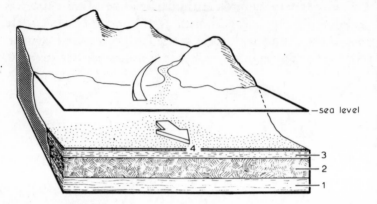

Fig 40 Successive layers of rock represent different periods of deposition. Clay, mud, and sand are washed from the land by river action and deposited in layers on the sea bed. Layers 1–3 represent progressively younger beds, while layer 4 is still being deposited

The formation of vast thicknesses of rock must have taken many many years, and could not be attributed to a single catastrophic event. Finally he concluded that, because fossil men are never found in any but the most recent rocks, most fossils represented life as it must have been long before the advent of man.

. Gradually the story of life on earth was pieced together. In 1801, Jean Baptiste Lamarck (1744–1829) wrote, speculating that 'fossilised individuals were the ancestors of still living and since altered species', while Diderot (1713–84) had already made the remarkable suggestion that change could have produced all living animals from a single common ancestor.

The ideas of Leonardo, Steno, Guettard. Diderot, and Lamarck had shown that evolutionary change could explain the fossil record and the diversity of animals living today. And all this was known by 1800. Yet it took another hundred years and the genius of Darwin to convince many people of the idea.

The reason for this is not difficult to find; it was that in western Europe the vast majority of people adhered to a view of life that was both colourful and easy to understand. Animals had simply been created:

'God then created the great sea-monsters and all living creatures that move and swarm in the waters, according to their kind, and every kind of bird...'

Fossils were the remains of animals drowned in a single catastrophic flood:

'And every living substance was destroyed which was upon the face of the ground, both man, and cattle, and the creeping things, and the fowl of the heaven; and they were destroyed from the face of the earth...'

This view of life was not only held by official spokesmen of the Christian Church. In 1715 a Swiss, Johann Scheuchzer,

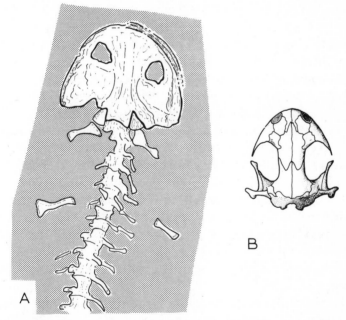

Fig 41 *Homo diluvii testis*, A, was thought to be the remains of a man drowned in the Flood. Cuvier realised that it was a fossil salamander by comparing it with living forms B. B is the skull of *Salamandra atra* viewed from above. Not drawn to scale

described *Homo diluvii testis*—a witness of the Flood (fig 41A). The fossil was that of a skull, badly crushed but with the eye sockets clearly preserved.

The strongest support for the idea that animals were created came from William Paley who published his *Natural Theology* in 1802. In it, he described many complex organs such as the vertebrate eye, and argued that they were so beautifully designed and so perfectly adapted to the animals' needs and environment, that they could only be the result of conscious design; design therefore, by God.

The Church was such a powerful force that even the methodical Steno, who had realised the significance of the 'tongue stones', abandoned his ideas. When he was almost fifty years old, he had planned to write down the results of his paleontological and geological research, but he never completed the task. He became a priest and spent the last years of his life in poverty in northern Germany.

Guettard, whose ideas so contradicted the literal interpretation of Genesis, was forced by the theological professors of the Sorbonne to publish a repudiation of his beliefs.

And all this happened in spite of the fact that Leonardo had effectively demolished the theory that Noah's flood could have been responsible for the fossils of Lombardy. A flood, he pointed out, could certainly carry remains of land animals to the sea, but could not have been responsible for the presence of marine forms so far inland. In spite of the fact, also, that the great French anatomist Georges Cuvier (1769–1832) was able to show *Homo diluvii testis* for what it really was—a giant salamander—by comparing it with closely related living forms (fig 41B).

It is curious, and sad, that a deeper understanding of life was delayed for so long. Sad too, because the conflict between biology and the Church was so unnecessary, for few Christians today would claim that the Bible is in any way negated just because it can no longer be read literally and unthinkingly.

6

Variation

The rocks of the English Lower Liassic are about 180 million years old and contain large numbers of a fossil called *Gryphaea*. *Gryphaea* is related to the oyster but differs from that species by having a variably coiled shell. Some specimens have a quarter of a whorl while others have about one and a half whorls. Between these two extremes there is a continuous series of variations, four examples of which are shown in fig 42.

A collection of *Gryphaea* made from a single layer of rock is called a population sample. If it is large enough, it is reasonable to suppose that the various forms in the sample are represented in the same proportions as existed in the population

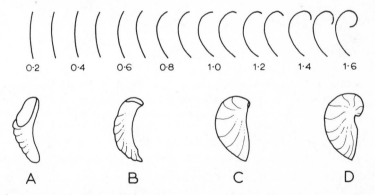

Fig 42 *Gryphaea* is a fossil relative of the oyster. Its shell is variously coiled having between 0·2 and 1·6 whorls. Samples of the various forms are shown here, A–D

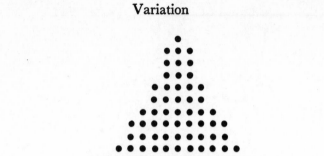

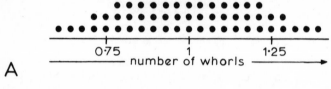

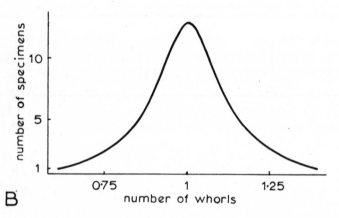

Fig 43 A population sample of *Gryphaea*. In A, each dot represents a specimen, and it can be seen that most specimens have about one whorl. The same information is expressed in the population curve in B

that lived while the rock was being deposited. A population sample is shown in fig 43A in which each dot represents a specimen. There are 109 specimens (in practice a sample is usually larger); some have less than 0·75 of a whorl and others more than 1·25 of a whorl, but most have about one whorl. The population as a whole can be represented by a curve (fig

74

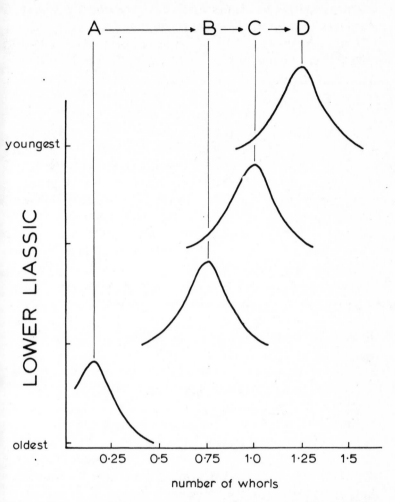

Fig 44 Four population curves representing samples of *Gryphaea* taken from successively younger rocks of the Lower Liassic. They show that the most common variants at the beginning of the Lower Liassic had less than 0·25 of a whorl, while those at the end of the Lower Liassic had 1·25 whorls. The most abundant variant in each population (A–D), corresponds to types A–D in fig 42. Based on a drawing from Swinnerton, *Outlines of Palaeontology*, 1947

43B) which shows how much diversity is present, and in which the peak indicates the most abundant types.

If similar curves are drawn for population samples taken from different horizons (ie, levels) in the Lower Liassic, we find that there is a gradual shift in the position of their peaks as one progresses from older to younger rocks (fig 44). The peak of the oldest population sample indicates that the majority of specimens have less than 0·25 of a whorl, the youngest population mainly consists of specimens with 1·25 whorls, while populations of intermediate age consist mainly of specimens with a single whorl. There was therefore a gradual change from type A to type D (fig 42) during the Lower Liassic, but this change was not absolute, it was rather one of emphasis. That is, type C became increasingly common in Lower Liassic populations of *Gryphaea* until it eventually became the dominant form.

In Chapter 4, I mentioned that it was an oversimplification to regard change as taking place between individual animals and plants. So, by applying the lesson learnt from *Gryphaea*, we can now refine our concepts of simple and divergent change.

Change is the result of an alteration in the relative number of a particular variant in a population.

Earlier we symbolised simple change from a white circle to a black circle as in fig 45A. Imagine now a population of white circles (bottom row of fig 45B). At some point a new variation occurs, a black circle; its effect on the population is at first negligible but, if its relative numbers grow, it will eventually replace all the original white circles. Simple change from a population of white circles to one of black circles will have taken place.

In fig 46A, the simplified symbol for divergent change is shown, a white circle produces two new forms. Figure 46B shows what this change is like in terms of white circles. At first their relative numbers do not change, but, as a result of

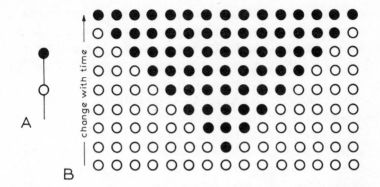

Fig 45 Simple change from white to black can be expressed as in A. In terms of a population, this change is represented by a gradual increase of black variants in successive generations B

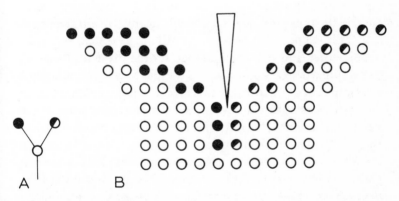

Fig 46 Divergent change expressed simply in A, and in terms of a population in B. In B, two variants, black and half black, occur in an early generation but do not greatly affect the population until some factor divides the population. After this division, simple change occurs in each of the new populations, but in one case it is towards a predominance of black individuals and in the other, towards half black individuals

77

some internal or external factor, the original population is divided into two parts. In one part, increase in the relative numbers of black circles produces change in one direction, while in the other, increase in the relative numbers of another variant produces change in another direction.

We see now why variation in a population is so important. Without it, change is impossible.

Gryphaea is not an isolated example. If any population is studied closely enough, innumerable variations are found among its members. Every human being is unique, as every successful burglar acknowledges when he wears gloves, and it is seldom difficult to recognise a friend, even in a crowded street. Variation in animals other than man is, of course, more difficult to spot, but this is because of our lack of training. If, as is normal among westerners, we find it difficult to distinguish between two Chinamen, we must expect to look very closely indeed to find variation in a flock of starlings.

There are two causes of variation in populations, environmental and genetic. By environmental, I mean any factor outside the organism, and by genetic, factors within the organism. This is not a very precise definition as we shall see in the next chapter, but it is good enough for the present discussion.

Every gardener knows that a simple way of producing a large number of identical plants is to divide a single root stock. The plants resulting from this splitting are the same as one another if they are grown under the same conditions (fig 47A). If, however, they are planted in a variety of situations, some dry, others damp, some in acid soil, others in chalk and so on, plants displaying a variety of form will be produced (fig 47B). Because we know that the various plants originate from the same stock, we may safely deduce that the variation observed is caused by the different environments.

Another example is illustrated by my own family, members of which have emigrated to different parts of the world during the past fifty years. Today, as a result of environmental influ-

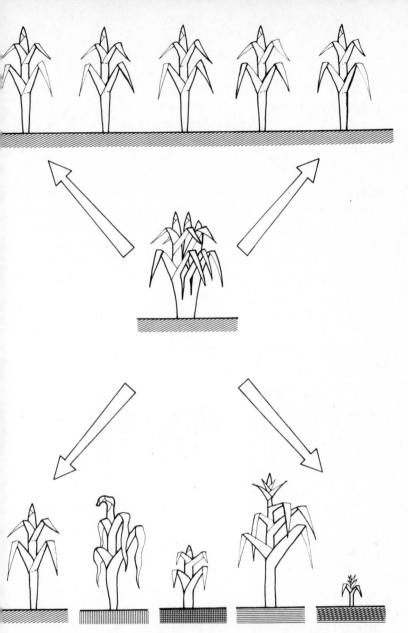

Fig 47 Variation caused by the environment. If cuttings from a plant are grown under identical conditions, a series of identical plants results, A. If these same cuttings are grown under various conditions a variety of plants is produced, B

ence, my various uncles, aunts, and cousins, speak a variety of languages, all different from my own.

Genetic factors produce variation in a population even when the environment is stable, so patterns caused by genetic variation are not always obviously correlated to environmental conditions. This is true of human blood groups, and the distribution of one, group A, is spread unevenly throughout the world, being most common in parts of Canada, Scandinavia, and Australasia, and almost non-existent in South America (fig 48).

Another example of genetic variation is particularly interesting because it concerns variation in behaviour rather than of structure. The stork *Ciconia*, lives during the summer in Europe and during the winter in South Africa, migrating

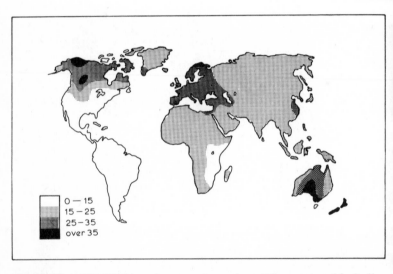

Fig 48 The distribution of a genetic variant. The geographical distribution of blood group A in human populations. The figures represent the percentage of sex cells produced in a given area that carry the gene responsible for blood group A compared to those carrying genes responsible for other groups. Based on a drawing from Mourant, *The Distribution of the Human Blood Groups*, 1954

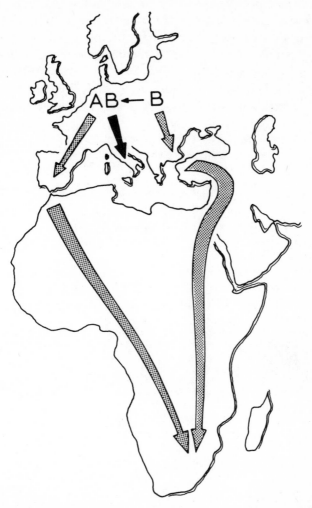

Fig 49 Variation caused by genetic factors in the stork *Ciconia*. Genetic variation has produced two stork populations, A and B. Population A migrates to South Africa via the western route while B migrates along the eastern route. If an individual is taken from population B and released from western Europe (where population A lives), it migrates as though it were still starting its journey in eastern Europe (black arrow)

between the two regions each year. Genetic variation has produced two groups, A and B. Members of group A live in western Europe and begin their migration by flying in a south-westerly direction, while members of group B live in eastern Europe and fly in a south-easterly direction. By doing this, each group avoids crossing the Alps during its migration south. That these two groups are genetically determined was

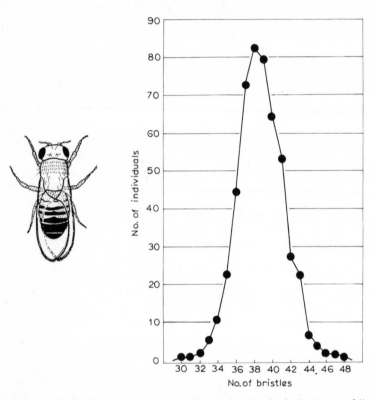

Fig 50 Variation caused by genetic factors in the fruit fly *Drosophila* (left). In a population of 511 individuals, most flies had 38 bristles on their bellies, but variants with 30 to 48 were also recorded. This variation occurred even though the flies were bred under identical environmental conditions

82

demonstrated by the following experiment (fig 49). A number of young storks were taken from group B, transported to western Europe, and kept in captivity until all the birds in group A had migrated. Instead of now taking the south-westerly route when released, as might have been expected, the young storks flew in a south-easterly direction in spite of the obstacles created by the mountains of the Alps. In other words, they obeyed genetic instructions rather than responding to their changed environment.

In the examples discussed so far, genetic factors produced different groups within populations, but in some cases more continuous variation can be produced. In an experiment, nine generations of a fruit fly called *Drosophila melanogaster* (fig 50) were bred in identical environments. In the resulting population of 511 flies, a number of variations were present. One that could be easily studied concerned the number of bristles on each fly's belly, and it was found that, with respect to this character, there was continuous variation of between 30 and 48 bristles, although the typical members of the population had about 38 (fig 50).

The amount of variation caused by the environment then, depends upon the diversity of environmental conditions, while genetic factors are independent of the environment and either produce different groups in a population or more continuous variation. What must be emphasised now, is that the examples given so far are unusual, for variation is usually the result of interaction between both environmental and genetic factors. The height of a man is determined by his genetic constitution and by his environment—the amount of food available to him, the presence in his diet of certain minerals, and so on. As a result, height in man is very variable and any population shows continuous variation with respect to this character. This is demonstrated by fig 51, in which the variation in height in a population sample of over eight thousand adult men is shown to be continuous between 57 and 78 inches (145cm and 198cm).

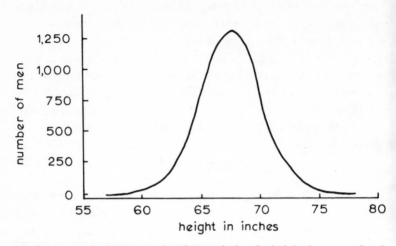

Fig 51 A population curve showing variation in height in a sample of 8,500 adult men. Such variation is caused by both genetic and environmental factors

Compare fig 51 with fig 43B (p74). Both curves are very similar, and it is likely that the variation in the amount of coiling in *Gryphaea* is the result of interaction between genetic and environmental factors.

One last point must be mentioned. In fig 52A, a variant (the black square) caused by genetic factors or by genetic factors modified by environmental ones, occurs in a population of

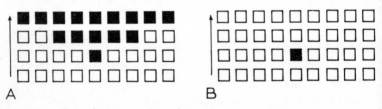

Fig 52 A variant (black square) caused by genetic factors can increase in successive generations of a population, thus effecting a change in the population. One caused by environmental factors cannot spread unless the environment in which the entire population changes, B

84

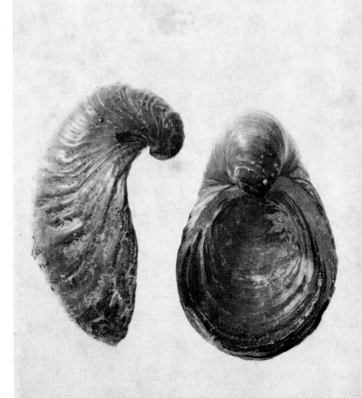

Page 85 (right)
A specimen of *Gryphaea* viewed from the side and the front. The coiling of the shell is clearly seen in the side view; *(above)* block of limestone from Staffordshire, England, deposited in a shallow sea some 400 million years ago. As the limestone was formed, remains of animals that lived in the sea were entombed and their skeletons preserved giving us a detailed picture of a great variety of organisms

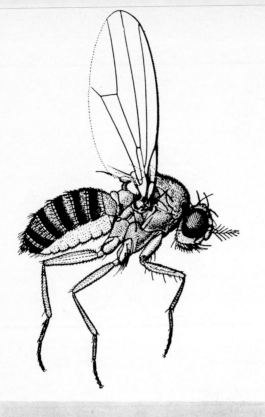

Page 86 (left) Droso-
phila melanogaster,
the fruit fly. Study of
variation in its anatom
led to an understand-
ing of how animals
change and the causes
of such changes;
(below) chromosomes
taken from a single
human cell. Careful
examination of this
picture reveals that
there are 46 in all, and
that they can be
arranged in 23 similar
pairs. Each pair is
called a homologous
pair and is composed
of a chromosome
derived from each
parent

white squares. Such a variation can be passed to succeeding generations (each horizontal row in the figure represents a generation), and there is the possibility that increase in its relative numbers could effect a change in the population. A variant caused by environmental factors on the other hand, cannot be passed on. It therefore cannot spread throughout a population and cannot effect any long-term change (fig 52B).

At some stage during the evolutionary history of the giraffe,

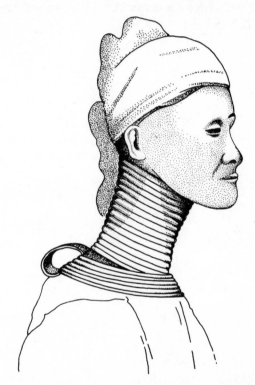

Fig 53 A giraffe woman from Burma. Her neck is long because it is affected by environmental factors—the rings. Such an acquired characteristic cannot be passed from one generation to another, and her children would have normal necks if they were not subjected to the same environmental factors

individuals with long necks were produced by genetic factors. Over many generations these variants spread throughout the giraffe population because they were able to take advantage of food that was out of reach of their fellows. A progressively longer neck evolved until it is seen in an extreme condition today. A long neck is also produced in human women from parts of Burma, but in this case the cause is environmental, the women being forced to wear rings around their necks (fig 53). Because this variant is not genetic in origin, Burmese children are never born with long necks, they only acquire them if they too are forced to wear rings.

7

Growth and Development

We have seen that change in animal populations is possible only if variation occurs and that, for reasons summarised in fig 52, this variation must be of a kind that can be passed from one generation to the next. How then is such variation produced? The brief answer is, by varied control of the growth and development of individuals exercised by structures called chromosomes. To understand what this means, we must first look at the process of growth and development and then at the ways in which chromosomes may affect this process. I will start by describing the development of a young vertebrate (fig 54 A–F), but the general remarks that follow apply to all animals in which sexual reproduction is usual.

Development begins with a single cell, the egg, which is produced by the female. Compared to other cells, the egg is large because it contains yolk which is the only source of food for the young embyro. Usually, the egg cell does not begin to divide until fertilisation has taken place. This involves the union of the egg with a much smaller cell produced by the male called a sperm cell, an event which takes place either inside or outside the female's body, A. The cell resulting from fertilisation is called a zygote. This divides, B, until a mass composed of small cells is produced, C. The exact shape of this mass, the blastula, is determined by the amount of yolk that was originally present in the egg cell. After repeated cell division, the blastula resembles a hollow ball, D. Events in the

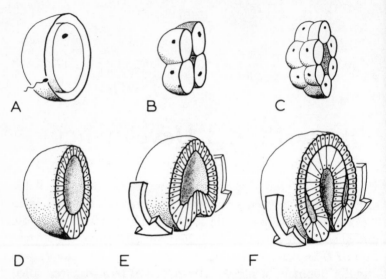

Fig 54 The early stages in the life of a vertebrate involve fertilisation during which male and female sex cells are united. Further development involves the multiplication of cells and the differentiation of layers of cells. After fertilisation A, the zygote cell divides B–C, to form a hollow ball called a blastula, D. This ball is partly turned inside out, E, until it has a wall composed of a double layer of cells, F–the gastrula

next stage of development are similar to those that would occur if you squeezed a tennis ball in an attempt to turn it inside out, E. This process is called gastrulation, and it transforms the blastula into a gastrula, which is still hollow, but has a double wall, F. Between these two walls, a third layer of cells usually grows, and it is from these three layers that all the various tissues and organs of the body are later derived.

The main points to be learnt from this extremely brief description are these: development involves multiplication of cells, so that, in man, a single zygote weighing about a 20 millionth of an ounce (about a millionth of a gramme) produces an adult of perhaps 150 pounds (68 kilogrammes); also, gradual differentiation of these cells takes place, for while the cells of the blastula are similar to one another, there is a great deal of difference in shape and function between say, muscle, nerve, and liver cells. The third point is so obvious that it might easily be missed. A zygote produced by egg and sperm cells from, for example, monkey parents, will always produce another monkey, never any other animal.

Another point about development is illustrated by a consideration of a later stage in the growth of one particular organ, and that is, that the process of cell differentiation can

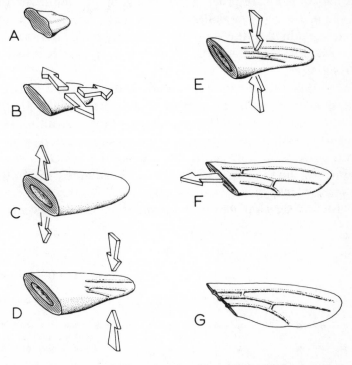

Fig 55 Development of the wing in *Drosophila*. The wing bud A grows as a result of cell division B, and later, a cavity forms within its tissues, C. Flattening of the wing begins at the tip D and continues E, until the fluids inside the cavity are expelled, F. During stages D–F wing veins are formed that give the adult wing G rigidity. Based on a drawing from Waddington, *Principles of Embryology*, 1956

be divided into a definite sequence of events. In fig 55 A–G, the development of the wing of a fly called *Drosophila* has been summarised. At first, the wing consists of a bud made up of cells derived from the outer layer of the gastrula, A. This bud grows as a result of normal cell division, B, but later expansion is due to separation of its upper and lower layers, C. Further development results in the flattening of the wing, which begins at the tip, D, and moves towards the base, E. At certain points, the two layers do not meet, so that veins seen in the adult fly are produced. As the wing is compressed, the tissue of the wing becomes more compact, and fluids which were contained in the wing-bud cavity are forced out, F. The final stages of development produce a wing that is thin but strong, G.

To emphasise the uniqueness of organic growth it is useful to compare it with the growth of a crystal (fig 56). In the correct solution, a crystal will grow according to definite laws, but its growth is no more than simple enlargement of a basic three-dimensional shape, in this case a cube, which is determined by the properties of the elements contained within the crystal. Hence there is no process of differentiation or of division of development into a sequence of discrete events as

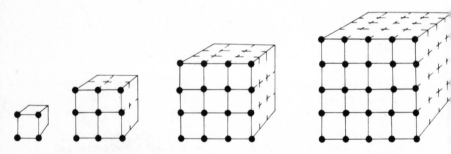

Fig 56 Organic growth is more complex than the growth of a crystal. This diagram shows that a crystal grows by the addition of material in an unchanging pattern. Compare this to the processes shown in figs 54 and 55

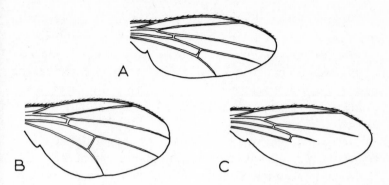

Fig 57 The normal wing of *Drosophila*, A, compared with 'broad', B, and 'veinlet', C, wings. For an explanation of how variants B and C are produced, see text

there is in organic growth. There is no evidence of complex organisation.

As I have said, the organisation of growth and development of animals is the responsibility of structures called chromosomes.

I chose to describe wing development in *Drosophila*, because many variations of that organ are known. Two of these are compared with the normal wing in fig 57. One variant is called 'broad' (fig 57 B) and the other, 'veinlet' (fig 57 C). Knowing a little about wing development, we can now guess how these variants have been produced. In the case of the 'broad' wing, cell division at an early stage of wing-bud growth (fig 55 B) has taken place more rapidly across the bud than along the bud. This uneven growth produces an unusually broad wing-bud, the characteristic seen later in the adult wing. The variation called 'veinlet' is different from the normal wing because its veins do not meet the hind edge of the wing. This is the result of late development of veins while the wing-bud was being compressed (fig 55 D and E). It will be remembered that in the normal wing, veins are first developed at the wing tip.

These examples show that variation in the shape and form of an organ such as a wing is the result of some change in one or more of the events that comprise its development. Sometimes, a single change at an early stage of development has the most profound effect on the form of the adult organ, while change at a later stage may have correspondingly slighter effects.

Two individuals in a population may appear to differ with respect to only a single character, but, more commonly, differences can be seen in a large number of characters. In the fruit fly, *Drosophila*, variation is known to occur in the size and colour of the eyes, the shape of the wings, the number of bristles present on the body, the shape, size, and colour of the body as a whole, and so on. During the early part of this century, T. H. Morgan (1866–1945) carried out numerous breeding experiments with a species of the fruit fly called *Drosophila melanogaster*, and he found that characters seen in adult fruit flies did not occur randomly in successive generations but that, when a particular variation occurred in an individual it was generally accompanied by some other variation of another character. For example, the wing called 'vestigial' (in which the wing is very incompletely developed), very commonly occurred in flies that had unusually dark bodies. These character variations were said to be linked. After a time it became apparent that all the variations seen in *Drosophila melanogaster* could be linked together into four groups. This was significant, because in every body cell of *Drosophila melanogaster* there are four pairs of chromosomes. The variations seen in the maize plant can be linked together into ten groups and, once again, the same number of chromosome pairs are present in its cells. There is a suggestion then, that chromosomes are somehow associated with the occurrence of variation in individuals.

What are chromosomes and how do they affect development and cause variation? Chromosomes are tiny thread-like struc-

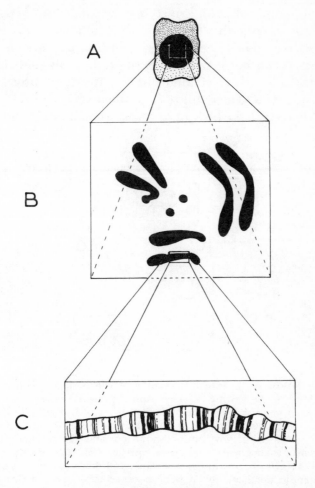

Fig 58 An animal or plant cell A contains a nucleus, within which the chromosomes are found. These are arranged in pairs, B, and each is seen to be banded if sufficiently enlarged, C

tures that occur in the cells of animals and plants (fig 58A). In *Drosophila* there are eight chromosomes, they occur in pairs, and the members of each pair are, with one exception, identical in shape and length, although members of different pairs may be quite different (fig 58B). If part of a chromosome is sufficiently enlarged, it is seen to be banded (fig 58C), and one can imagine that this is the result of a large number of parts arranged in a linear sequence. Such an arrangement does, in fact, occur, but it can only be appreciated by studying the molecular structure of the main constituent of chromosome material, a substance called deoxyribonucleic-acid (DNA).

Every cell in the body of an animal contains the same num-

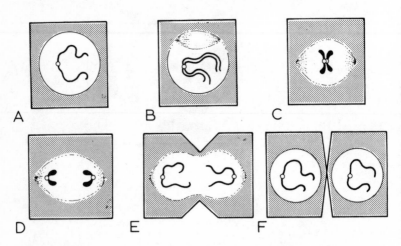

Fig 59 Mitosis, or normal cell division. A cell is shown with a single chromosome A. During a stage called prophase, the chromosome divides to produce two chromatids that are joined together at a single point; meanwhile a spindle is formed at the edge of the nucleus, B. This spindle then moves into the nucleus and the chromatids, now thickened, assume a position on its equator. This stage is called metaphase, C. The chromatids then separate and move to opposite ends of the spindle, anaphase, D. At stage E, called telophase, the chromatids become elongate again, and the cell begins to divide. This division is complete at stage F so that two cells have been produced from one, each containing the same chromosome material

ber of chromosomes. This is because, as we have seen, the development of an animal involves the repeated division of cells, and an essential part of this process is the duplication of chromosome material, so that when a cell divides to produce two new cells, the new cells contain the same chromosomes which were contained in the original cell. The way in which this duplication takes place has been summarised in fig 59, but the details of this complex process (called mitosis) need not concern us here. All that is significant to the present discussion is that the chromosomes found in the cells of an old man are exactly the same as those which occurred in the zygote from which he grew, and that their presence made this co-ordinated growth and development possible.

DNA, the chemical from which chromosomes are made, is an enormous molecule composed of two chains of sugar molecules linked by phosphate groups (fig 60). Between each pair of opposing sugar molecules there are two molecules called bases. If the phosphate-linked sugar molecules are likened to the sides of a ladder, these bases may be visualised as the rungs. There are, however, two important differences between

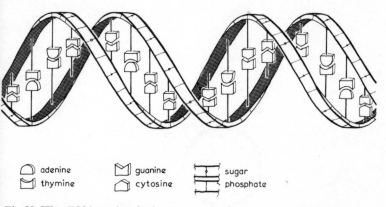

adenine   guanine   sugar
thymine   cytosine   phosphate

Fig 60 The DNA molecule is composed of two helix spirals of sugar and phosphate molecules linked by pairs of bases, adenine, thymine, guanine, and cytosine

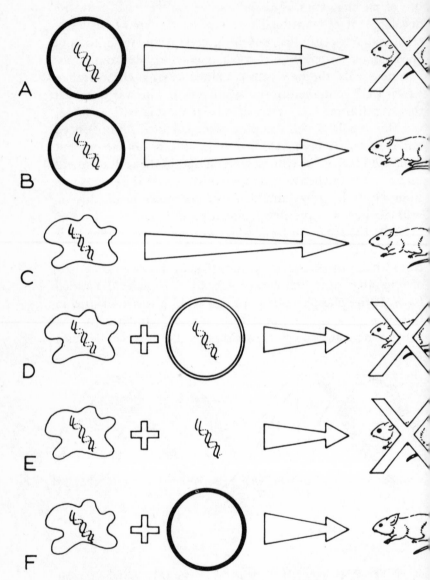

A

B

C

D

E

F

Fig 61 Experiments with a bacterium called *Diplococcus pneumoniae*. For explanation, see text

98

DNA and a ladder, because in DNA there are many thousands of rungs, and the sides are twisted in the shape of a spiral. The bases that form the rungs of DNA are of four kinds; adenine, thymine, guanine, and cytosine, and they are always arranged in pairs: adenine joined to thymine, and guanine to cytosine.

The importance of DNA can hardly be exaggerated, for an organism can be changed completely by changing the DNA in its cells. This has actually been achieved in experiments carried out on a bacterium that causes pneumonia, *Diplococcus pneumoniae*. *Diplococcus* occurs in two forms, one, the virulent strain, is contained within a microscopic spherical capsule, while a non-virulent strain resembles a tiny blob of jelly. If the virulent strain is injected into a mouse or rabbit, the animal very soon dies of pneumonia (fig 61A). The same strain, if it is first killed by heat treatment, is similar to the non-virulent strain in that it has no adverse effects (fig 61 B and C). In the 1920s a bacteriologist called Griffith made a surprising discovery. He found that a mouse, injected with non-virulent *Diplococcus* and the bodies of the dead virulent strain, contracted pneumonia (fig 61D). The significance of this discovery was realised years later when it was found that the non-virulent strain caused pneumonia if it was injected together with DNA extracted from the virulent strain (fig 61E). In other words, it was demonstrated that the effects of *Diplococcus* are not due to the presence or absence of the surrounding capsule, as might be supposed, but to the DNA it contains. Conclusive proof of this assertion comes from a final experiment (fig 61F), in which a mouse survives after having been injected with the non-virulent strain together with empty capsules from the virulent strain.

The nature of an organism then, is governed largely by the DNA in its chromosomes. The way in which this remarkable molecule affects the development and final form of organisms has been the object of much research during recent years. Gradually, the following picture has taken shape.

Growth and Development

Organic processes depend upon the presence of chemicals called proteins which are composed of chains of smaller molecules, called amino-acids. Although there may be as many as a million different proteins in the body of one man, there are only twenty amino-acids. This apparent discrepancy is explained by the fact that a protein is defined by the exact order in which its constituent amino-acids are linked together. DNA is responsible for the production of proteins in the cell, and it is the order in which the bases are arranged in the DNA molecule that determines the order in which amino-acids are strung together. As the DNA molecule is enormously long, it is able to control the production of many different proteins.

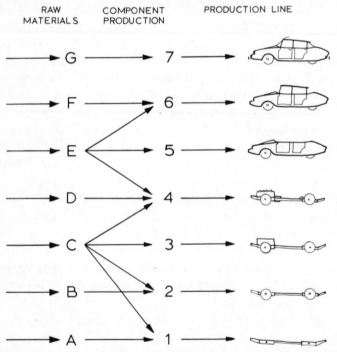

Fig 62 The body is composed of proteins which are themselves made up from amino-acids. In the same way, raw materials A–G are used to make components 1–7 which are assembled to produce a motor car

The way in which DNA determines the development of an animal or plant can be likened to the manufacture of a motor car (fig 62). Raw materials A–G, are brought into a factory where they are used to produce components 1–7, car bodies, engines, wheels, and so on. Obviously many different raw materials are used to produce certain components. At the production line, the various components are joined together until the final product is complete. With each stage of an industrial process such as car manufacture, the units produced gain in complexity. Raw materials are chosen for specific properties, steel, rubber, and glass for example, but the components which are produced from these materials have even more distinctive properties; an engine is able to perform tasks

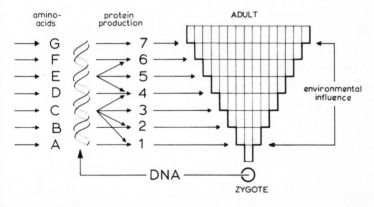

Fig 63 The assembly of an animal. Amino-acids A–G are the raw materials of the body. These are organised by DNA into proteins 1–7, the body's component parts. Because DNA is reproduced every time cells in the body multiply, it is always present to supervise protein production. The final form of an animal is determined both by the instructions given by its DNA, and by external environmental influences

quite different to those carried out by a gear-box or a chassis. The final product, a car, is even more complex than any of its component parts for, although embodying them all, as a whole it performs a function which none of its components could perform alone.

We have seen that development of an adult animal from a zygote involves multiplication and differentiation of cells and that development takes place in a definite sequence of events. This sequence is not very dissimilar to that which takes place on a car production line. An animal is built of amino-acids which we may liken to the raw materials of the previous example, and these are organised by different parts of the DNA molecule into proteins, the body's component parts (fig 63). Each protein has an essential role in development although other factors such as the environment play a part. If part of the DNA code is changed, different proteins are produced and the developmental process is affected, resulting in some variation in the young individual. A change in the DNA code in *Drosophila* may cause a new protein to be manufactured that affects cell division in the wing bud, so that a variant such as 'broad wing' may be produced. In the same way, a different engine could be fitted to a car thus changing its performance ability.

Page 103 Flowers of the wild rose have five petals. By selecting particular variants, a process called artificial selection, horticulturists have been able to breed roses with far more petals. This process is not significantly different from natural selection which generates new animal and plant types under natural conditions

Page 104 Charles Darwin (1809–82), the biologist who first elucidated a theory of evolution by natural selection. Although considerably refined since his day, the theory is still known as the Darwinian theory of evolution

8

Genetic Variation

Molecules are small but animals are much larger. For this simple reason, it is far easier to see the effects of DNA change than it is to investigate the change itself. The study of animal and plant variation began well over a century ago, long before DNA was discovered, so it is not surprising that the gene was postulated. The gene was visualised as a hereditary factor causing variation; we now have a more precise model, for we

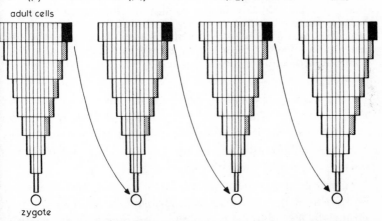

Fig 64 A single cell, the zygote, divides to produce the cells of an adult animal. At an early stage, germ cells (stippled) are produced, and these eventually produce gametes (black), the cells responsible for succeeding generations. Because chromosomes are reproduced with every cell division, chromosome material is passed from the parent generation P to succeeding generations F1–F3

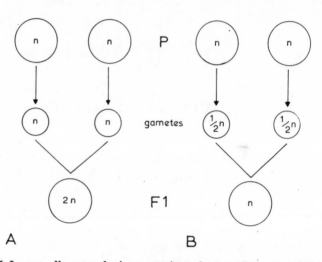

Fig 65 In sexually reproducing organisms, both males and females, P, contribute a number of chromosomes to the next generation F1. If there were no reduction in the number of chromosomes when the gametes are produced, the number of chromosomes present in one generation would be twice the number present in the parent generation, A. This does not occur because there is such a reduction, B. This reduction occurs during meiosis (see fig 69)

can define it as a part of the DNA chain that is responsible for the production of a protein that has a well-defined effect on the organism. Sometimes the effect of a single gene is obvious but, more commonly, whole suites of genes play a part in the development of a particular character.

It will be remembered that a factor causing variation must be inherited before it can have any lasting effect on a population (fig 52, p00). We can now see how this is possible. Chromosome material is duplicated every time a new cell is produced in the tissues of an animal or plant. At an early stage of development, a number of cells are produced which make future generations possible. These are the germ cells or, more specifically, the cells which are destined to produce gametes, so that the cells of a young animal contain chromosomes derived from its parents (fig 64).

If, in sexually reproducing organisms, the gametes contained the same number of chromosomes as the body cells of the parents, the number of chromosomes would double with each generation (fig 65A). This does not in fact occur, because when the gametes are produced there is a reduction in the number of chromosomes (fig 65B). The process of cell division that produces gametes is therefore different to normal cell division and is called meiosis. As a result of meiosis, not only is the number of chromosomes in members of a species kept constant, but any individual inherits half its chromosomes from one parent and half from the other (fig 66 A–C).

I have already mentioned that in the cells of any animal or plant the chromosomes occur in pairs; these are called homologous pairs. Each member of a homologous pair is derived from a different parent and carries the same number of genes arranged in linear fashion like so many beads on a thread. Genes that occupy similar positions on homologous chromosomes are called alleles, and they function together by influencing one particular biochemical event in the processes of body development and maintenance. Often, one of a pair of alleles is more powerful than the other in its effects, and is therefore called the dominant gene. Alleles are usually symbolised by an initial letter; the dominant gene by a capital and the other, the recessive gene, by a lowercase letter (fig 66D).

To illustrate the action of alleles, let us take for an example the four o'clock, *Mirabilis jalapa*, a flower that is either red, white, or pink. These colours are the result of the presence of proteins that are produced during the development of the flower, proteins that are determined by a single pair of alleles. One gene R, produces a red flower; its allele r, a white flower; and a combination of both, a pink flower. A pink flower is

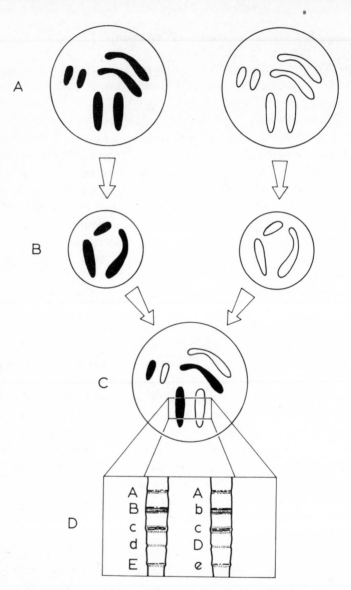

Fig 66 The inheritance of paternal and maternal chromosomes. In A, chromosomes from male (black) and female (white) germ cells are shown to include three homologous pairs. Production of gametes, B, results in cells with only three chromosomes, one from each pair. After fertilisation, the resulting zygote cell C, has three pairs of chromosomes, the same number as in the parent cells. An enlargement of parts of a homologous pair of chromosomes, D, shows that alleles are found on corresponding positions on these chromosomes

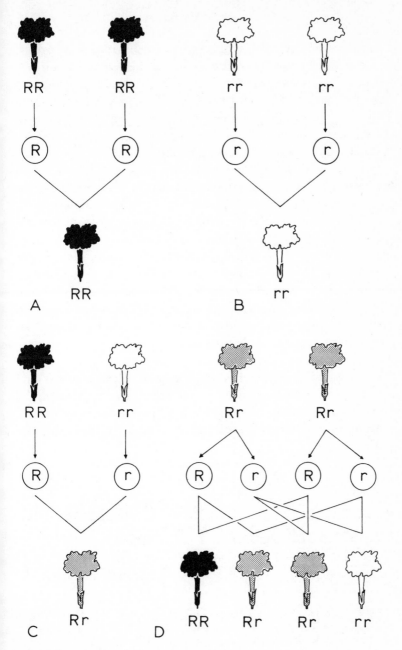

Fig 67 The inheritance of flower colour in the four o'clock, *Mirabilis jalapa*. For explanation, see text

possible because there is incomplete dominance of R over r. The cells of a red flower then, contain the genes RR and it is said to be homozygous for that particular gene. As a result of meiosis, such a flower can only produce gametes containing the gene R, so that, if a red flowering plant is crossed with another red plant, all the offspring will also be red (fig 67A). In the same way, two plants homozygous for the gene r, and therefore with white flowers, can only produce white flowered offspring if crossed (fig 67B). If, however, a red flowering plant is crossed with a white plant, all the offspring have pink flowers because the next generation inherit both the alleles R and r (fig 67C). Such plants are said to be heterozygous. Finally, if two pink flowering plants are crossed, each parent will produce gametes containing either one of the alleles, which then combine as in fig 67D to produce a generation of offspring that have either red, pink, or white flowers. Because the most likely combination of genes is R with r, about half these offspring will have pink flowers.

The example just described shows that variation in a character occurs even if a single pair of alleles controls that character. Usually, however, characters are determined by groups of alleles in which there is scope for much more variation. Even if only two alleles affect a character, they can come

	AB	Ab	aB	ab
AB	AABB	AABb	AaBB	AaBb
Ab	AABb	AAbb	AaBb	Aabb
aB	AaBB	AaBb	aaBB	aaBb
ab	AaBb	Aabb	aaBb	aabb

Fig 68 The possible combinations of two genes A and B, and their alleles a and b. There are nine: AABB, AAbb, aaBB, aabb (once); AABb, AaBB, Aabb, aaBb (twice), and AaBb (four times)

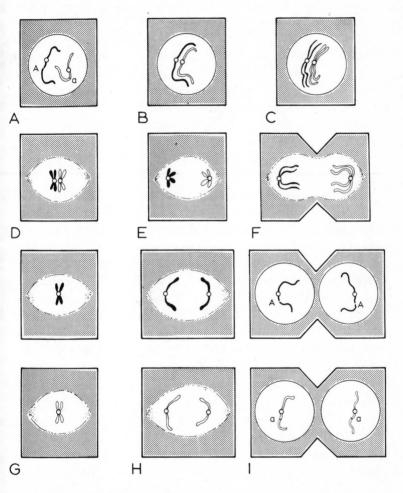

Fig 69 Meiosis, or reduction division. The details of cell division in which gametes are produced which contain half the normal number of chromosomes. For explanation, see text

111

together to form nine combinations (fig 68). The number of actual variations seen in such a population will depend on how the different combinations of alleles react with one another during development. Imagine now a character determined by twenty alleles. The different possible combinations of genes are astronomical, and almost continuous variation of the character affected can result. We have already seen an example of such variation, the number of bristles on the belly of *Drosophila melanogaster* (fig 50, p000).

So far, I have discussed the kind of variation that results from the fact that sexually produced organisms inherit genetic material from both parents. But there are other sources of genetic variation, and also mechanisms that limit variation. Neither of these topics can be discussed without first considering the process of cell division called meiosis.

The sequence of events that comprises meiosis is illustrated in fig 69. It takes place in the ovary and testis and, in both cases, germ cells with the normal number of chromosomes (the diploid condition), produce gametes, eggs and sperm, that contain half the normal number of chromosomes (the haploid condition). We begin then, with a diploid cell in which, for the sake of simplicity, only two chromosomes are assumed to be present, one inherited from the mother (white) and one from the father (black) A. At the beginning of meiosis, the chromosomes come to lie alongside one another that is, homologous chromosomes form pairs, B. They then divide longitudinally along most of their length to produce what are known as chromatids, C. There follows a process that is of cardinal importance because it causes variation, but let us ignore this for a moment, for it demands closer attention later. The chromatids become short and thick, and arrange themselves on a spindle which at this stage replaces the nucleus in the cell, D. They then move to opposite ends of this spindle, E, and become elongate again, F. At this latter stage, the cell divides so that two cells are produced. There follows a second cell division in which the chromatids become separated. Once again, a spindle forms within each cell and the chromatids arrange themselves on its equator, G. They then split, and the two halves, now normal chromosomes, move to opposite ends of the spindle, H. Thus at the second division, four gametes are produced, each containing half the normal number of chromosomes.

This outline of meiosis shows how, as a result of reduction of chromosome number in the gametes, cells are produced which contain one of any pair of alleles. So that if the alleles A and a are found on corresponding loci on the chromosomes in fig 69A, half the gametes produced will carry A and the other half a. This same phenomenon, called segregation, made possible the combinations illustrated in fig 67.

We must now consider the segregation of two pairs of alleles, say Aa and Bb. If each of these alleles occur on chromosomes that are not homologous, there is no problem, gametes containing the combinations AB, Ab, aB, and ab are produced

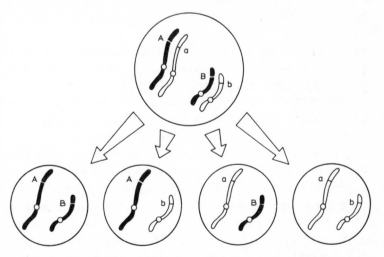

Fig 70 The production of four gametes from a single germ cell in which gene combinations AB, Ab, aB, and ab are produced

113

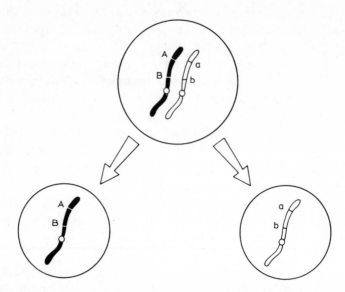

Fig 71 The production of two gametes from a single germ cell in which gene combinations AB and ab are produced. Combinations Ab and aB are not produced because genes AB and ab are linked, that is, they occur on the same chromosome

(fig 70). But if these alleles are located on homologous chromosomes, fewer combinations are possible and only gametes containing AB and ab are produced (fig 71). The genes A and B and a and b are said to be linked, and their tendency not to be separated during meiosis is responsible for the occurrence of linked characters mentioned in the last chapter.

The association of linked genes is not, however, absolute. For a phenomenon called 'crossing over' sometimes occurs. Crossing over makes it possible for genes which are not linked to come together during meiosis. To understand how this mechanism works we must look more closely at the stage of meiosis called diplotene (fig 69C). This is done in fig 72 A–E. To begin with, let us assume that alleles Aa and Bb occur on homologous chromosomes, A. At the onset of diplotene the

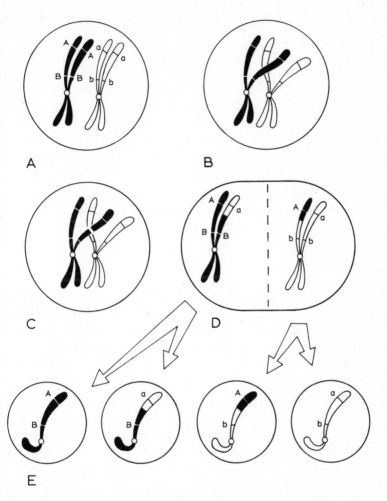

Fig 72 The production of four gametes, E, from a single germ cell A. The situation is identical to that seen in fig 71 except that crossing over occurs enabling the production of gametes with combinations Ab and aB as well as those with AB and ab. For fuller explanation, see text

chromosomes, which are divided into chromatids, become intertwined, B. In the event of a break in two chromatids, C, rejoining can result in the exchange of part of one chromatid with that of another, D. The germ cell then divides twice as outlined in fig 69, with the result that, if the break occurred between the loci of Aa and Bb, gametes can be produced in which all possible combinations of the alleles are represented, AB, aB, Ab, and ab, E. In effect, the result is exactly the same as when alleles Aa and Bb were located on chromosomes that were not homologues.

All the genetic variation discussed so far is the result of combination in the zygote of genes derived from two parents. But there are other kinds of variation that are caused by actual changes, or mutations, in the gene itself, or by re-arrangement of genes on the chromosome or by chromosome multiplication.

The effects of mutations are often said to be extreme, but this is probably not true. What is certain is that mutations with obvious effects are most easily spotted by zoologists. It has also been a common misconception that mutations always confer disadvantages on the organism possessing them. Once again, the statement is only partly true, for many mutations are very small and it is difficult to see their effects whether they be for good or ill. Mutations are random events, but it is likely that only a very small percentage produce advantageous effects. An animal or plant is an intricate piece of machinery that is always in a state of delicate equilibrium. If you have ever built a castle with playing cards, it will be apparent that, after a certain point, the addition of a single card is more likely to disrupt the entire edifice than it is to make the castle bigger. Exactly the same is true of the effects of mutations on organisms.

Mutations appear to occur at different rates in different organisms. For example, a mutation that is responsible for the production of resistance to the drug streptomycin in the bacterium *Escherichia* appears to occur in one cell in 1,000 million in each generation. A more frequent rate has been recorded in fruit flies in which some mutations occur in one cell in 100,000 in each generation. In fact, rates of mutation are probably quite variable, even in a single animal. They are affected by X-rays and other forms of radiation, and an increase in radioactivity will speed up the rate at which mutations occur. This last fact has been shown by analysis of mutation rates in survivors of the bombing of Hiroshima (fig 73). The frequency of the mutation causing some kinds of leukemia is much higher in survivors who were close to the

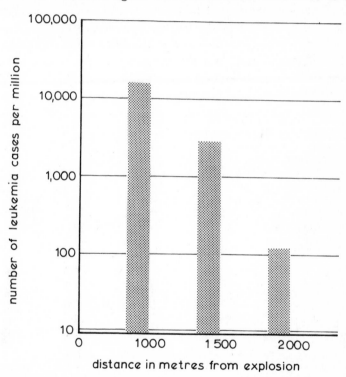

Fig 73 Rates of occurrence of leukemia during the period 1950–57, caused by chromosome damage in survivors of the bombing of Hiroshima. People who were unfortunate enough to be close to the explosion received larger doses of radiation

explosion centre, and who therefore received a much higher dose of radiation than those who were a mile or more away.

One of the commonest mutations to occur in animals is that causing albinism. In the fruit fly, *Drosophila*, the eyes are normally black, a colour produced by the mixing of red and brown pigments. Several genes are responsible for these pigments, and mutations causing any of these genes to be ineffective during development produce a number of varieties. If the brown pigment is not produced, pink-eyed individuals result; if the red pigment is not produced, brown eyes result; and if neither pigment is produced, the eyes are colourless.

As well as change in the biochemical structure of a gene, re-arrangement of existing genes can have profound effects which cause variation in a population. Such re-arrangements are of two kinds, inversions and translocations. Both are the result of breaks in the chromosome being followed by rejoining in new positions, and both cause re-arrangement in the order of genes on the chromosomes involved (fig 74).

Finally, chromosome multiplication can sometimes occur.

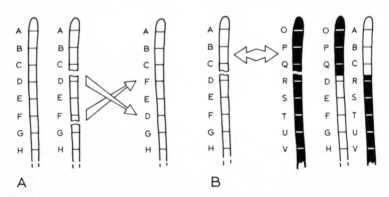

Fig 74 The re-arrangement of gene sequences as a result of chromosome damage. In inversions, A, a section of a chromosome is broken and rejoined in a reversed position. In translocation, B, fragments from homologous chromosomes are exchanged

As outlined earlier, meiosis ensures that half the normal number of chromosomes are present in the gamete cells. Sometimes, however, this mechanism breaks down and, while the resulting offspring usually die while young or are infertile, sometimes a completely new variety is produced. Such new forms contain a simple multiple of the haploid number of chromosomes possessed by their ancestors, and are called triploids or tetraploids, depending on whether this multiple is three or four. Multiplication of chromosome numbers often produces a new species and, although it is extremely rare among animals, it has been responsible for the formation of as many as a third of the known species of flowering plants. Two examples of the latter may be mentioned here. The normal diploid paeonies are found in southern Spain, Crete, and parts of Asia Minor, but various members of this group have produced tetraploid

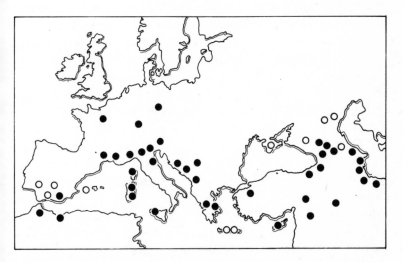

Fig 75 Paeonies with the normal number of chromosomes (white circles) have given rise to a number of tetraploid forms (with twice the normal number of chromosomes, black circles) which extend throughout the Mediterranean region. Based on a drawing from de Beer, *Atlas of Evolution*, 1964

forms which have increased the range of the group and have spread throughout Europe and Asia Minor (fig 75). The second example is of great economic importance, for many of the species of wheat that are cultivated by man are tetraploid and even hexaploid forms of wild strains.

Genes, chromosomes, meiosis, and mutations, all these seem far removed from the topics of earlier chapters, so it seems advisable to summarise what has been discussed so far, so that we may appreciate what ground has been covered, and what remains to be discussed. We began with the fact that animals are similar to one another and can, as a result, be classified into groups. These groups are members of a hierarchy in the system of classification, and tend to be limited to discrete geographical areas or to particular periods of the earth's geological history. The distribution in space and time of animal groups is only explained if we assume change to have taken place. Change is the result of the introduction and spread in population of new variants. Such variants are produced by new combinations of genes and by gene mutations— random biochemical changes in a molecule called DNA.

So far so good, but there is one obvious discrepancy in our story so far. The kind of change that can be seen to have occurred during the past, is well ordered and with apparent direction; the taeniodonts are a good example (fig 26, p47). But the variation caused by new combinations of genes or by mutations is a random affair and, as we have seen, often produces rather unsuitable characters, flies without wings, men with leukemia, and so on. How is it then, that animals with the potential to change in a number of ways, in fact change in a way that enables them to carry on living successfully? The answer forms the subject of the next four chapters.

9

Darwin's Theory

In the last two chapters we have seen that inheritable variation is caused by random biochemical changes. Let us now consider what would happen if these events were the only factors in the mechanism causing change in populations. This is best done diagrammatically (fig 76). Starting with a square, we see that by 'reproducing' offspring that display completely random variation, a very varied population results. With a second generation, this variation becomes even greater, and no two individuals are alike.

Such a mechanism obviously does not operate in animals. A horse produces offspring that show some variation, but they

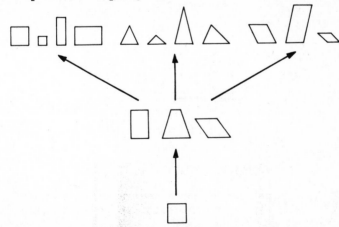

Fig 76 Production of a random variety of shapes from a square

are still horses nevertheless. We know from the fossil record that animals have changed, but we also know that such change is very gradual and is also in a sense directed. Within a particular lineage, change of the kinds represented in figs 29 and 30 (pp55 and 56) may take place, but such change is quite different from the continuosly diverging change indicated in fig 76.

Some zoologists have tried to explain orientated change by suggesting that mutations are directed rather than random. There are many examples in the paleontological record of animals in which a particular feature shows gradual change in one direction. An example is the Irish elk, *Megaloceros,* in

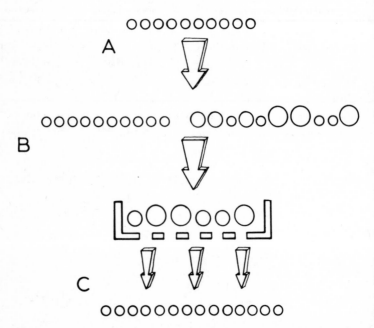

Fig 77 The operation of a selective mechanism. To a collection of marbles A, others of various sizes are added, B. To ensure that the final collection contains marbles of constant size, a sieve can be used to select marbles of one particular size

which the antlers increased in size until they were incredibly large. The suggestion is, that in this particular line, the only mutations to occur were those that produced even larger antlers. This theory of evolution, sometimes called ortho-genesis, is false, for there is no evidence that mutations are ever directed.

Mutations then, are random. But we have seen that random variation alone cannot account for the kind of change that we know occurs. Obviously other mechanisms operate.

Imagine a set of ten similar marbles (fig 77A). To this set, let us add ten more, which are of various sizes (fig 77B). This additional set represents the variation produced by sexual reproduction of a new generation. Now let us consider what we must do to keep our collection constant with respect to size. One way is suggested in fig 77C; a sieve is used that allows marbles from our original collection, and any others in the additional set that are similar, to pass through, while other marbles are retained. We would then have a new collection of marbles of similar size and could throw the rest away.

Another problem. How can we change our collection of marbles, replacing small ones with larger ones? Once again, we add a new generation of marbles to our original set. We then use the sieve as before, but this time retain those marbles that are caught in the sieve and throw the others away. The size of the marbles in our collection would, of course, depend upon the size of the pores in the sieve. So, in spite of having no control over the size of marbles added in each case to our original collections, we can determine the size of the marbles in our final collections. We do this, however, only by satisfying two conditions: we must be prepared to sacrifice marbles that are not required; and we must use a selection mechanism, in this case a sieve.

Now let us consider animal populations to see whether the experiments with marbles have any relevance. Firstly, is there a wastage of unsuitable members of successive generations?

A female halibut can lay 5 million eggs in a year, other fish as many as a million, and many insects over a thousand. At the other extreme, birds may produce clutches of between 2 and 10 eggs every year of their adult life which, in the case of a sparrow, may be fifteen years. Even if we suppose that an animal has only two offspring during its entire reproductive life, that animal could, in theory, be responsible for the production of over a thousand descendants after only ten gener-

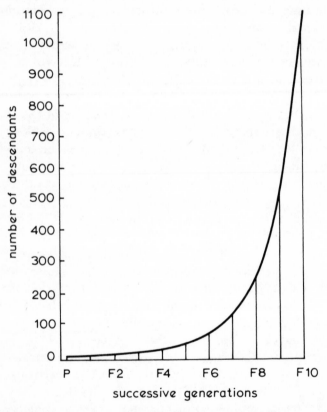

Fig 78 Even if an animal has only two offspring during its entire life, it could, in theory, be responsible for over a thousand descendants after ten generations

ations (fig 78). In other words, animals are able to reproduce at a rate that is far greater than is needed simply to maintain existing numbers in any population.

Now, it is apparent that the increase in the number of animals in any population, which in theory is possible, is never in fact achieved. This is because a large number of members of any generation die before they reach maturity. In fig 79, a hypothetical population is shown, A, the seven dots could represent seven or seven thousand individuals. Such a group may produce over twice as many eggs, in this case 17, but a number will never become fertilised, B. This number varies considerably, but is especially high for animals in which external fertilisation takes place. It is because the chances of fertilisation in the open sea are so slight, that fish such as the

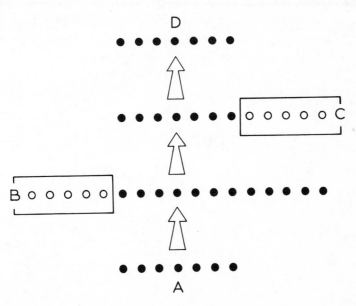

Fig 79 A population A produces a large number of eggs but, of these, some fail to become fertilised, B, while others produce young that are killed before maturity, C. Those remaining, D, are the effective next generation

halibut produce so many eggs. The remaining fertilised eggs begin to grow but, before they produce mature offspring, many will be killed by predators and disease, C. Mortality rates are especially high during the early stages of an animal's life, after it has severed physical contact with its mother (in the case of placental mammals) and before it has fully learnt to fend for itself. The number of offspring that do eventually reach maturity, D, is similar to the number in the parent generation.

Of course, fig 79 is an idealised example, it is a model, not the record of an actual population. What usually happens is that populations vary in size from generation to generation, and are seen to be stable only after they have been studied for a considerable period of time. This has actually been done in the case of hares in Canada, where, since the mid nineteenth century, they have been trapped for their fur. The records of numbers caught, kept by the Hudson's Bay Company, are likely to be a good reflection of the actual numbers of hares present. These records show that the hare population fluctu-

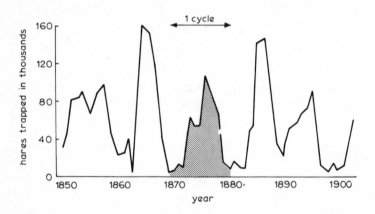

Fig 80 Fluctuations in the hare population of Canada are reflected in records of their capture. It can be seen that the population expands rapidly, then declines, a cycle that is repeated roughly every ten years. Based on a drawing from Young, *The Life of Vertebrates*, 1950

ated considerably, reaching a peak every nine or ten years (fig 80). At the beginning of each cycle, the number of mature offspring is greater than that in the parent generation so, for a time, the population rises sharply. After a time, factors such as food shortage and an increase in the numbers of predators, lead to a sharp decline in the population, although over the fifty-year period shown, the population as a whole does not undergo any change of great significance.

It is clear then that a combination of two facts, that animals can potentially produce vast numbers of offspring and that populations are more or less stable over long periods, leads to the conclusion that indeed high mortality rates are normal in animal populations. So far, natural circumstances appear to parallel our game of marbles.

Now to the second point. Is there a selection mechanism that determines which members of a population do and do not survive? For it is obvious that, if the pruning of a population is random, successive generations would still reflect the random variations generated by gene mutations.

The answer is yes, there is a sieve: it is a mechanism called natural selection and it was elucidated by Charles Darwin (1809–82). Natural selection simply means that individuals that are not suited to survive in certain environments do not survive. This definition will be refined later but, for the moment, it is good enough to allow us to consider some examples of its operation.

Let us first see how natural selection works in an environment that does not change. Such an environment is found in deep sea regions where a lamp shell called *Lingula* lives. Fossil members of the genus *Lingula* are known from rocks that are 500 million years old, and they still survive today. Palaeozoic populations of *Lingula* must have produced a number of variants including those with, for example, differently shaped shells (fig 81). These variants were not suited to the deep sea environment and so perished, leaving no offspring.

The only ones to survive were those that were similar to their ancestors. This process of selection for the same shell shape must have continued for millions of generations, for the shell is exactly the same today as it was during the Silurian period or even earlier.

For selection in a changing environment, let us take a second look at *Gryphaea*, the oyster discussed at the beginning of Chapter 6. It will be remembered that the shell of this fossil

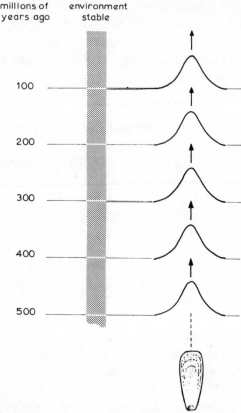

Fig 81 The deep sea environment of the lamp shell *Lingula* has remained stable for some 500 million years. As a result, the shape of the shell in this animal has remained unchanged. The curves on the right represent population samples (compare with fig 44, p75)

128

became increasingly coiled during the Lower Liassic. If we now suppose that during this period the sea bed changed from one that was rocky to one that was muddy, we can guess why the change took place (fig 82). On rocky substrata, the flat *Gryphaea* could live with its upper valve facing upwards, and take in food and water through a gap between this valve and its shell. When mud became more common, a coiled shell was of advantage, because it lifted the upper valve clear of the sea bottom, and prevented mud from entering the cavity in which the *Gryphaea* lived.

We have already seen that the change in *Gryphaea* was one of emphasis in successive populations (fig 44, p75). This is the inevitable result of selection against individuals that do not have a degree of coiling suitable for the environment at any given time.

The two examples just cited are similar in that in both cases, the environmental factors causing stability or change in population are of a physical nature. But there are other examples in which change is due to selection by another animal. The horse, for example, is a herbivore with no very efficient defence against would-be predators. When it is attacked it relies on its speed to escape potential danger. Now, during the history of one lineage of horses there has been a change from an animal the size of a dog to that which we see today, an increase in size by a factor of 4. As a result, the weight of the horse has increased by a factor of 64 ($4 \times 4 \times 4$). This increase

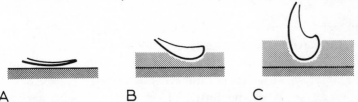

A   B   C

Fig 82 Changes in the environment and shell form in *Gryphaea*. In A, a flat *Gryphaea* rests on a rocky substratum; with increasing amounts of mud, B–C, a curved shell enables *Gryphaea* to keep the gap between its valve and shell clear of the mud

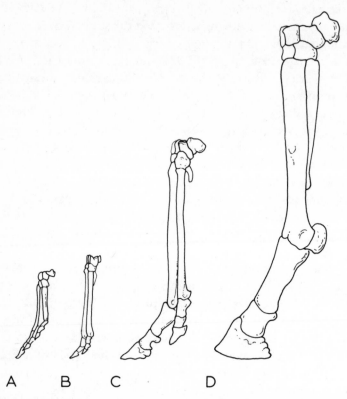

Fig 83 Change of size and structure of the front foot in fossil horses. A *Hyracotherium*, B *Mesohippus*, C *Merychippus*, D *Equus*. Drawn to scale. From Simpson, *Horses*, 1961. With permission of Doubleday and Co

in weight would have lowered the horse's speed, which in turn would have made it more susceptible to attacks by predators. However, variations occurred that enabled some individuals to maintain a high speed, and obviously these were the ones to survive. Study of the fossil record has revealed a series of changes that accompanied increase in size, changes that are best seen in the structure of the limb bones (fig 83). The most important ones involved an increase in overall length, greater rigidity enabling the springing gait of the galloping racehorse, a reduction in the number of toes and a relative increase in the length of the lower leg.

One final example of selection processes at work is that of the peppered moth *Biston betularia* whose British populations have changed during the past century in response to both physical and biological factors. The normal *Biston* is speckled grey and brown, colours that make it almost invisible when viewed against the lichen-covered bark on which it rests during the day. Prior to the 1850s almost all peppered moths were camouflaged in this way except that occasionally a mutation occurred that produced an almost black individual. These latter individuals probably never survived for long, because they were so conspicuous and were easily spotted by birds such

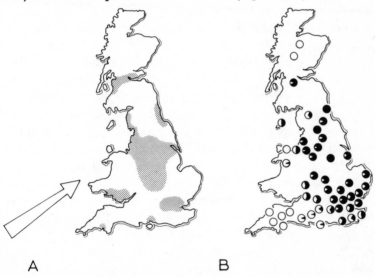

A   B

Fig 84 Correlation between the location of industrial areas of the British Isles, A, and the occurrence of the dark form of the peppered moth *Biston betularia*, B. Note that populations in which most individuals are dark (predominantly black circles) occur in or to the northeast of industrial regions. For explanation, see text. Based on a drawing from de Beer, *Atlas of Evolution*, 1964 (after Kettlewell)

131

as the spotted flycatcher, the nuthatch, and yellowhammer, all of which feed on *Biston*.

The middle part of the last century saw the development of large industrial complexes mainly in central and south-eastern England (fig 84A). The factories belched smoke that drifted north-eastwards under the influence of prevailing south-westerly winds, with the result that most of the lichen on trees in the eastern half of Britain was killed, and tree trunks became blackened. In this new environment, not only did the normally coloured peppered moths become easy prey for birds, but the dark variety benefited because it was now less conspicuous. As a result of these improved chances of survival, the dark form became increasingly common until now, it is by far the dominant form in many places (fig 84B). Only in unpolluted regions such as northern Scotland, North Wales, Devon, and Cornwall, do populations occur in which the normal form of *Biston* predominates.

The incredibly rapid change in *Biston* populations is due partly to the profound changes in its habitat during the last hundred years, partly to the efficiency with which birds select individuals that are not camouflaged, and partly to the fact that the mutation which produces a dark individual is dominant to the normal gene. The last reason is of interest because new mutations are usually recessive, that is, they must occur in the homozygous condition before their effects on the individual are apparent. Some four thousand years ago, most of Britain was covered with pine forests, a habitat in which natural selection would favour black forms of *Biston*. It is likely therefore, that normal populations of *Biston* were once black, that the light-coloured variety spread through its populations as the pine forests were replaced by deciduous trees, and that the change that has been observed during the last hundred years, is in fact a reversal back to the once normal condition.

Natural selection then, is the mechanism that operates auto-

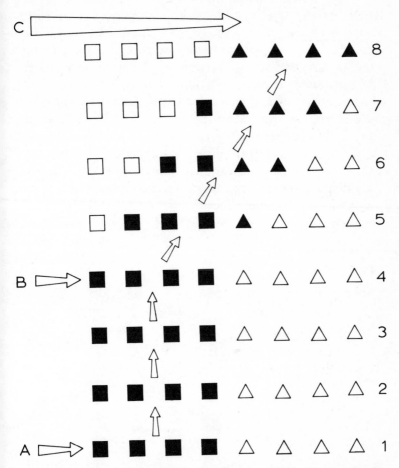

Fig 85 Change in a population caused by natural selection. In the initial population, two varieties occur, squares and triangles. At period A, selection occurs in favour of squares, and triangles are eliminated in each generation. At period B, a change in the environment causes selection against squares and favours triangular variants. As a result, in subsequent generations (5–7), triangles become increasingly common, until, at period C, they have replaced squares entirely

matically in populations and allows those members best suited to the environment to survive at the expense of less suited members. It is the mechanism by which the products of random variation are trimmed to ensure that populations remain if conditions are stable, and change if conditions change (fig 85). The idea that natural selection means survival of the fittest is, however, not really precise. We have seen earlier, that mere survival of an individual has no long-term effects on a population. For a population to be affected, the survivors must reproduce their kind. Selection then, is not just for survival, but for the ability to reproduce; and evolutionary change does not result from survival of certain variants, but from the fact that some variants contribute more successful individuals to successive generations, than do others.

10

Geological Change

Change in the environment is one of the most important causal factors of evolution. Without it, populations would become perfectly adjusted and then possibly never change. The environment is defined partly by physical factors such as climate, temperature, water salinity, and so on, and partly by biological factors. In this chapter I want to discuss physical aspects of the environment, the ways in which they change, and the effects such changes have on evolutionary processes. Long-term changes of the physical environment are a major concern of the geologist and before discussing them, it is necessary to mention, albeit briefly, the geological time scale.

There is a more or less continuous fossil record preserved in rocks that are up to 600 million years old. Six hundred million is such a large number that it is difficult to imagine. If one year is represented by an inch, then 600 million years is equivalent to about 9,500 miles, which is almost the distance between London and Tokyo, and greater than that between New York and Moscow. On the same scale, the taeniodonts discussed in Chapter 3 existed for a period equivalent to about 400 miles, and man, broadly defined as an animal capable of making tools, to a distance of only nine miles. The last 600 million years of the earth's history has been sub-divided by geologists into eras, periods, and epochs, and these are shown in fig 86. Of course the earth's history extends beyond the Cambrian period, some very early rocks are

Eras	Periods	Epochs	time since beginning
CENOZOIC	QUATERNARY	Recent	today
		Pleistocene	4
	TERTIARY	Pliocene	11
		Miocene	25
		Oligocene	40
		Eocene	60
		Palaeocene	70
MESOZOIC	CRETACEOUS		135
	JURASSIC		180
	TRIASSIC		225
PALAEOZOIC	PERMIAN		270
	CARBONIFEROUS		350
	DEVONIAN		400
	SILURIAN		440
	ORDOVICIAN		500
	CAMBRIAN		600

Fig 86 The geological time scale. Figures represent millions of years

thought to be 3,000 million years old, but rocks older than Cambrian rarely contain fossils and therefore need not concern us here.

One of the most profound changes to have taken place during the last 600 million years is a change in the relative positions of the continents. The notion that continents have drifted about the surface of the earth is generally credited to the German geologist Wegener who, in 1915, published a great deal of evidence in support of the theory. In fact, the idea had already been suggested in 1858 by Snider who supposed

Page 137 Pictures of the peppered moth *Biston betularia* show the importance of colour variation as a factor affecting survival: under normal conditions, the lighter variant has more chances of survival because it is less conspicuous against a background of lichen covered bark *(left)*.

In polluted regions where lichen has been killed and the bark of trees has been affected by sooty deposits *(above)*, the dark variant is less conspicuous and therefore is less likely to attract the attention of predatory birds

Page 138 The tuatara, a lizard-like reptile, is the last survivor of a group that was common over 200 million years ago. Its survival today is due to its isolated home on a few islands off the coast of New Zealand

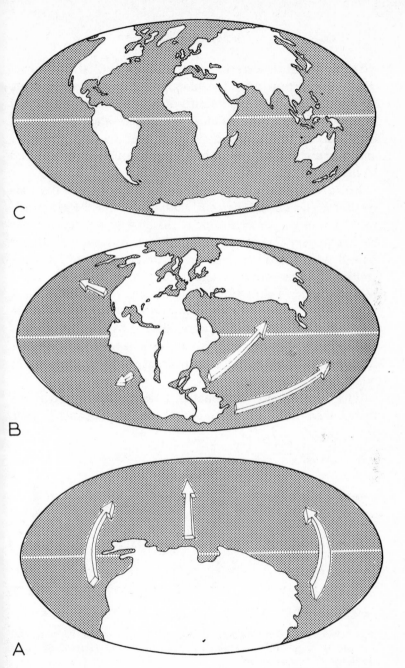

Fig 87 Movements of the continents during the past 400 million years.
A, pre-Carboniferous; B, Triassic; C, today

that the Americas were joined to Europe and Africa during the Carboniferous period. His evidence for this remarkable supposition was that fossil plants of the Carboniferous period from both sides of the Atlantic are largely identical. Today, the theory of continental drift is accepted by most geologists and a great deal of research during the past few years has strengthened their belief.

Broadly speaking, the theory states that prior to the Carboniferous period, all the continents of the world were joined to form a single supercontinent that lay in the south polar region (fig 87). During Permian times this supercontinent moved northwards and later began to fragment to form the continental masses that we know today (fig 87 B and C). These movements have affected animals and their evolution in two ways. Firstly, as the continents moved northwards, they came closer to the equator and so were affected by climatic changes and secondly, the fragmentation of land masses isolated groups of animals and led to a greater diversity of forms than would otherwise have occurred.

The best evidence for climatic change caused by northward drift is seen in the succession of rocks of Permian and Triassic

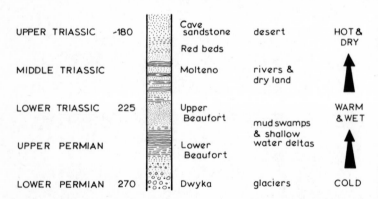

Fig 88 Different sediments deposited in Africa during the Permian and Triassic periods indicate changing conditions and climates. Figures represent millions of years

age in Africa (fig 88). During Lower Permian times, South Africa lay close to the South Pole, and rocks deposited at the time, those of the Dwyka formation, are typically glacial sediments that would be expected in a cold region. By the Upper Permian, some 45 million years later, the so called Beaufort beds were being deposited. These indicate that warmer conditions existed, for they consist of mudstones that were deposited in a swampy environment. Detailed analysis of these rocks shows that South Africa was then covered with mud swamps with occasional freshwater lakes, rivers, and deltas. During the next 40 million years, the period called the Triassic, Africa moved closer to the equator. Gradually the swamps dried out and sandstone rocks indicating hotter conditions were deposited. By Upper Triassic times some 180 million years ago, red deposits called the Red beds and Cave sandstone were deposited that are typical of hot desert conditions.

Study of the rocks of southern Africa then, shows that climatic conditions changed from very cold to very hot in the course of about 100 million years.

Apart from causing climatic changes, continental drift has been responsible for the isolation of faunas; we have already seen how the marsupial mammals paralleled the placentals as a result of their isolation on Australia when that continent drifted away from Africa and Asia (fig 37, p65). In this connection it is interesting to note, as did Snider, that before the continents drifted apart, some plants and animals occurred over what is now a large geographical area. The Permian fossil plant *Glossopteris* for example, is found in rocks of many regions in the southern hemisphere (fig 89A), but if its occurrence is plotted on a map in which the continents are reconstructed in their positions during the Permian period, its distribution is much more intelligible (fig 89B).

A second example is seen in a family of Triassic fishes called the Cleithrolepididae (fig 90). Two forms, *Cleithrolepidina*

141

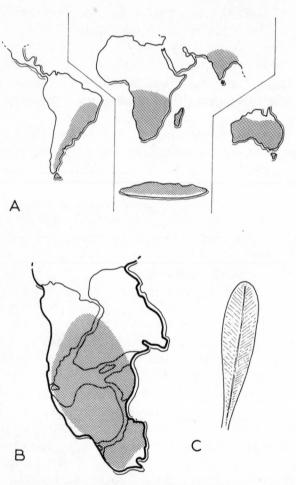

Fig 89 Fossil remains of the Permian plant *Glossopteris*, C, are found in a number of regions, A. If the continents are fitted together in their Permian position, the distribution of *Glossopteris* is seen to occupy a single area

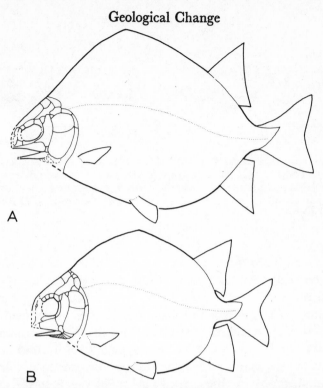

A

B

Fig 90 Two Triassic freshwater fish, *Cleithrolepidina* from South Africa A, and *Cleithrolepis* from Australia B, are extremely similar to one another. Their occurrence in widely separated parts of the world is explained by the fact that, during the Triassic, the Indian Ocean did not exist. Drawn three-quarters natural size. Based on drawings from Hutchinson, 'A Revision of the Redfieldiiform and Perleidiform fishes from the Triassic of Bekker's Kraal (South Africa) and Brookvale (New South Wales)', *Bulletin of the British Museum (Natural History), Geology* Vol 22 No 3, 1973

from the Lower Triassic of South Africa, and *Cleithrolepis* from the Middle Triassic of New South Wales, are remarkably similar to one another; so similar that they almost certainly have a close common ancestor. Now, the surprising fact about these fish is that they are both freshwater forms, and are unlikely therefore to have been able to cross the Indian Ocean. It follows then, that at the time when these fish were

143

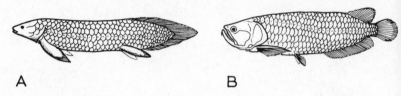

A   B

Fig 91 Survivors of once widespread groups are still found in Australia. The lungfish *Neoceratodus* A and an osteoglossid *Scleropages* B. Their ancestors reached Australia before the continent had drifted away from Africa. Not drawn to scale. From Norman, *A History of Fishes*, 1963

alive, there was no Indian Ocean to cross. There are only two truly freshwater fish living in Australia today, a lungfish called *Neoceratodus*, and an osteoglossid, *Scleropages* (fig 91), and both are survivors of groups that were once widespread. It is almost certain that the ancestors of both these fish, like *Cleithrolepis*, reached Australia before that continent drifted away from Africa. As a result of the paucity of freshwater fish in Australia, many river habitats have become available for colonisation by marine groups that could evolve forms which were able to withstand conditions of low salinity, and this in fact has happened.

Another great change in the physical conditions of the earth took place during the Pleistocene or Ice Age, about four million years ago. At that time, ice sheets spread from the North Pole until they covered much of North America, Europe, and the mountains of Asia (fig 92). There were four periods of glaciation separated by much warmer periods. Obviously, animal life was greatly affected by these extreme climatic changes. During the cold periods, animals normally associated with arctic conditions, the marmot, snow lemming and arctic fox, reindeer and musk-ox, were found as far south as the border between Germany and Switzerland, where their remains have been found in cave deposits. During the warm interglacial periods, the reverse situation occurred, and ani-

144

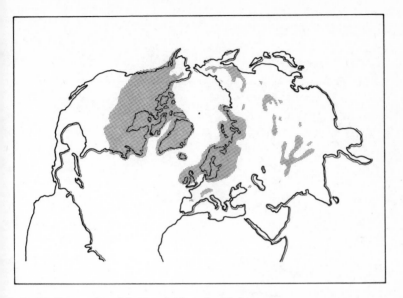

Fig 92 Areas covered by ice during the Pleistocene

mals typical of warmer climes, lions and hippopotamuses, spread as far north as Great Britain. These, however, were only changes in the distribution of animals, it is more diffi- cult to demonstrate evolutionary change that may have been caused by the uneven Pleistocene climates. This is probably because the extreme conditions lasted for a relatively short period. Nevertheless, some changes did occur, and we find furry relatives of otherwise naked forms during the Pleisto- cene such as the woolly rhinoceros and the mammoth.

Another effect of the Ice Age was to isolate various popula- tions of a single species. This occurred when the ice finally retreated from its position over the British Isles. Trout lived in streams which flowed from the melting edge of the ice sheet and, as the latter retreated at the end of the Ice Age, these trout moved northwards. Inevitably, trout were trapped in water-filled hollows of the underlying land surface, and in

145

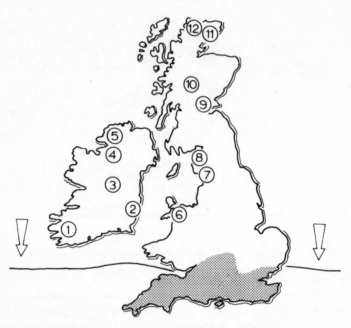

Fig 93 The British Isles was covered by ice during the Pleistocene except for its extreme southern part (stippled). As the ice receded, populations of trout were stranded in lakes and river systems and have since evolved to form distinct subspecies. 1 Cole's char, 2 blunt nosed Irish char, 3 Scharff's char, 4 Gray's char, 5 Trevelyan's char, 6 torgoch, 7 Windermere char, 8 Lonsdale's char, 9 Loch Leven trout, 10 Struan char, 11 Malloch's char, 12 large mouthed char

this way, populations of a single species were introduced into what are now the lakes of Ireland, North Wales, the Lake District, and Scotland. As the conditions in these lakes differ slightly, and the trout in each are isolated from one another, selection of different variants in each population occurred. Today a number of subspecies of trout survive in the British Isles, most of which are limited to a single lake or river system (fig 93). As yet, natural selection has produced only minor

146

differences between these populations, but in time these may well become increasingly developed and a number of new species will have been produced.

Geographical isolation not only encourages the evolution of new species, it also makes possible the survival of a species that otherwise would have become extinct. During the Triassic period very common animals in all parts of the world except North America were the rhynchosaurs, reptiles characterised by a beak on the upper jaw (fig 94A). These animals are now extinct except for a single relative, the tuatara or *Sphenodon* (fig 94B), that lives on a few islands off the coast of New Zealand. It is almost certain that *Sphenodon* would also be extinct, as a result of competition from other animals, if it had not been for the processes of erosion that produced these islands and provided an isolated home for this curious reptile.

Island formation, whether by erosion of a coastline or by volcanic action, almost always affects the evolution of certain animals and plants. Newcomers to newly formed islands often have a varied and new environment to colonise, and natural selection results in rapid change and diversification of the immigrant species. The Galapagos finches were a case in point (figs 25 and 36, pp45 and 64). Another feature of islands is that

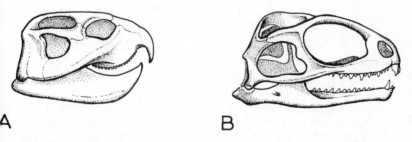

A   B

Fig 94 The skull of a rhynchosaur A, a member of a once common group that has a single surviving relative, the tuatara *Sphenodon* B. Not drawn to scale

they are often small, and almost always much smaller than corresponding habitats on the mainland. As a result, food is less abundant and smaller individuals of any species may well have definite advantages over larger members of the same species. It is not surprising then that many island species or populations are much smaller than their mainland counterparts; a pigmy form of the hippopotamus for example, lived on islands of the Mediterranean during the Pleistocene.

Let us now turn to changes in the physical environment caused by the processes of erosion and deposition. As water flows in streams and rivers to the oceans, it carries particles of weathered rock in the form of silt, mud, sand, and gravel. This material is eventually deposited as sediment on river and lake bottoms or on the sea bed. We have already seen the effects of the deposition of an increasing amount of sediment on Gryphaea (fig 82, p129).

A more complex example of evolutionary change correlated

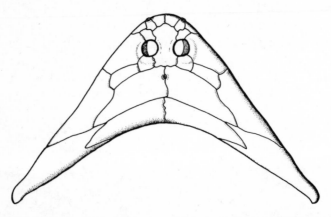

Fig 95 The skull of the extinct amphibian, *Diplocaulus*, viewed from above to show the horns whose function is unknown. Drawn one-quarter natural size. Based on a drawing from Beerbower, 'Morphology, Paleoecology, and Phylogeny of the Permo-Pennsylvanian Amphibian *Diploceraspis*', *Bulletin of the Museum of Comparative Zoology, at Harvard College* Vol 130 No 2, 1963

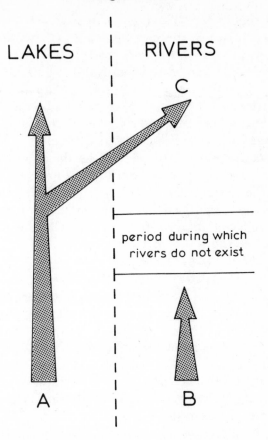

Fig 96 During early Permian times, two species of *Diplocaulus* A and B, lived in lakes and rivers respectively. After a dry period, during which the river species became extinct, species A underwent divergent change to produce a new species C that colonised rivers

with changing river and lake conditions concerns the now extinct amphibian *Diplocaulus* from the United States. *Diplocaulus* is easily distinguished from other fossil amphibians because its skull bears elongations or 'horns' (fig 95). The function of these horns is not certainly known, but it has been demonstrated that two species occurred during early

Permian times: one that is found in river and stream sediments and in which the horns point backwards to give the skull a streamlined shape; and another, from lake deposits, in which the horns are more laterally directed. Detailed analysis of the sediments deposited while *Diplocaulus* was alive shows that climatic changes were responsible for a period during which the streams and rivers dried up. Later, however, the streams and river re-established themselves. The history of *Diplocaulus* was greatly affected by these events (fig 96). At first, both species lived successfully, but with the disappearance of the streams and rivers came the extinction of the species with streamlined horns. When the dry period was over, varieties of the lake form migrated to the rivers and there evolved into a new species.

The examples just quoted show that evolutionary change can often be related to environmental changes which are geological in nature. However, the story in each case has been simplified to some extent, for, even though some process such as continental drift may be seen to be a causal factor affecting the evolution of an animal or group of animals, its effects are rarely direct. What usually happens is that a physical change occurs, and the whole balance of a community of animals and plants in a given area is upset. Evolutionary changes are simply adjustments made to restore this ecological balance. In other words, evolutionary change can only be appreciated if it is viewed in the light of all the conditions, physical and biological, that ultimately affect the life of an animal. The modifications and changes in the structure, physiology, and behaviour of an animal which enable that animal successfully to exploit its environment are called adaptations, and these are the subject of the next chapter.

Means to an End

In 1802, William Paley published his *Natural Theology* in which he catalogued a large number of adaptations. The intricate means by which different animals were able to exploit their environment impressed Paley so much that he saw them as evidence of a super-natural designer, just as the movement of a watch is evidence of the skills of a watchmaker. As proof of the powers of natural selection, many adaptations are no less impressive.

One is impressed by the adaptations that enable a bird to fly, not only because they are so numerous, but because they are seen to work in perfect harmony. Muscle, bone, feathers, and physiology are all modified to achieve a single end—flight. Even a brief analysis shows how many factors are involved. To begin with, the bones of the forelimb are greatly modified (fig 10, p23). The humerus is short and the bones of the hand reduced in number although those that are retained are elongated. There is little freedom of movement at the joints except at that between the wing and shoulder girdle. The power for the downstroke of the wing is provided by the pectoralis major muscle (the chicken's 'breast') which is supported by an ossified cartilage called the keel. To prevent collapse of the body when this muscle is contracted, the shoulder joints are braced apart by the coracoid and clavicle bones, the latter being fused together in a V, the furcula or wishbone (fig 97). The upstroke of the wing is affected by

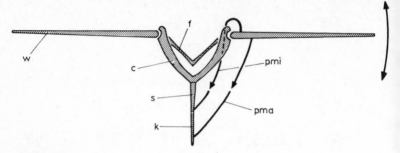

Fig 97 The skeleton of a bird seen from the front, c corocoid, f furcula, k keel, s sternum, w wing, pma pectoralis major muscle, pmi pectoralis minor muscle. For explanation see text

contraction of the pectoralis minor, a muscle that lies underneath the pectoralis major and which lifts the wing by means of a pulley system. Flight is only possible because the body is light, and the bird has a number of adaptations that save unnecessary weight. The bones are thin and often contain air sacs: where strength is needed, struts are used rather than thick solid bone. The body is streamlined to cut down wind resistance, and covered with feathers which act as flight surfaces and combine the properties of lightness with good insulation. The latter is necessary to keep the body temperature at the high level required by such an extremely active animal.

As I have said, one is impressed, not only by the large number of adaptations for flight in a bird (the list above is very incomplete), but by the way in which various parts are coordinated. Enlargement of the pectoral flight muscle is only possible if the keel is large enough to support it, and if the coracoids and clavicles are strong enough to take the extra strain. Fast flight is only possible if parts of the brain are enlarged to deal with the large number of rapid computations necessary for successful manoeuvrability in three dimensions, and so on.

If different animals adopt the same way of life they are likely to evolve similar adaptations. It is well known that the

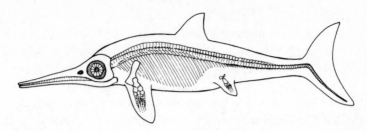

Fig 98 A ten foot long marine reptile from the Jurassic, *Ichthyosaurus*, had a similar body outline to that of many fishes

body shape of an extinct reptile, the ichthyosaur, has a similar outline to that of many fishes (fig 98). Differences of habit, on the other hand, lead to the evolution of various modifications of the same adaptive theme. The dorsal fin of fish, for example, functions as a hydrofoil to help maintain stability but performs many other functions as well (fig 99A–D). In fish such as the pike the dorsal fin lies in a position close to the tail and is an important swimming aid enabling short sudden bursts of speed A; in the stickleback the front fin rays have

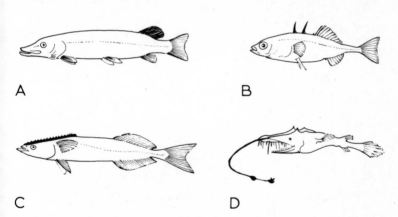

Fig 99 Modification of the dorsal fin in a variety of fishes. In the pike A, it acts as a locomotor organ, in the stickleback B it is a protective organ, in *Remora* C it is a sucker, while in the angler fish *Lasiognathus* D it is a luminous bait. Not drawn to scale

153

become protective spines B; while quite extraordinary modifications are seen in the remora and certain angler fish. *Remora* has a dorsal fin that is modified to form a sucker, C, enabling it to hitch free rides from larger animals such as sharks and whales. In the angler fish *Lasiognathus*, the front spine of the dorsal fin has been transformed into a rod from which a luminous bait is suspended to attract unwary prey, D.

Not surprisingly there are many examples of the evolution of specialised feeding methods in animals for under most conditions, an animal able to utilise a food source that is not available to other species, has a great advantage over its competitors. Thus the insects called bugs have mouth parts which enable them to feed on the blood of mammals or on sap from plants. They are able to do this because the mouth parts normally found in other insects, for example in the locust (fig 100A), have been modified to form a structure not

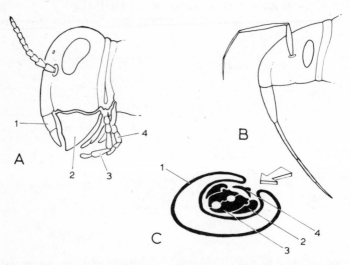

Fig 100 The mouthparts in the locust A, are modified to form a suctorial tube in some bugs B, as can be seen in the section C. 1 labrum, 2 mandible, 3 maxilla, 4 labium

Page 155 A leaf insect *(above)* and a stick insect *(right)*. Some insects are secure against predators because their bodies have come to resemble objects that are common in their environments

Page 156 The tree frog (above) has discs at the end of its digits which enable it to cling to smooth surfaces. A similar adaptation is seen in reptiles such as the lizards called geckos *(below)*. The evolution of similar adaptations in distantly related groups is called convergence

unlike a hypodermic needle which encloses two tubes, one for injecting saliva, and the other through which to suck food (fig 100B and C).

The teeth in different mammals, during the course of their evolution, have become adapted to a wide range of functions, some of which are illustrated in fig 101.

Adaptations are not seen only in the structure of organs; behaviour can also be adaptive. The archer fish is able quietly to approach its prey, an insect or similar morsel, and bring it to the water's surface and within reach of its jaws with a well-directed jet of water (fig 102). Another well-known behavioural adaptation has enabled blue tits to take advantage of the cream at the top of bottles of milk. The habit of pecking through milk bottle tops during the early hours of the morning is a commonplace activity among tits living in suburban areas, and serves to show how rapidly behavioural adaptations may become established in a population. Finally, birds such as the thrush commonly use stones as anvils on which to break open the shells of snails. I shall have more to say about behaviour and its relevance to evolution in the next chapter.

Meanwhile, to stress the variety of adaptations, let me mention some more examples, this time adaptation affording animals protection in different circumstances. Animals with distinctive shapes have often evolved body coloration that tends to break up their outline and makes them less conspicuous. Others change their colour each winter from brown to white in order to merge better with their surroundings, for example the arctic hare and fox. Yet more are able to change their colour very rapidly to suit prevailing conditions, as do the plaice and flounder. Coloration is not the only means of camouflage in the animal world; it is just as effective to resemble some commonplace object in the environment to be inconspicuous. Hence the sea dragon of Australian coastal waters has outgrowths that resemble seaweed (fig

157

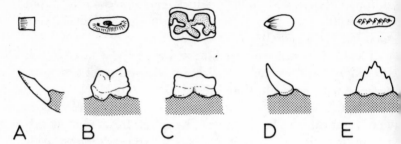

Fig 101 Teeth in different mammals modified for a variety of functions, seen in side and top view. A rodent (gnawing), B dog (cutting), C horse (grinding), D cat (piercing), E an extinct whale *Prozeuglodon* (chopping pieces of fish). Not drawn to scale

Fig 102 The archer fish, *Toxotes jaculator* is able to bring prey within its reach by aiming a jet of water. One quarter natural size. From Norman, *A History of Fishes*, 1963

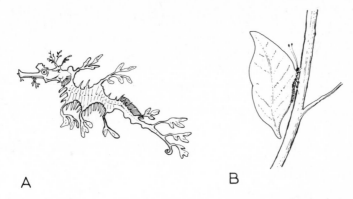

Fig 103 Animals that mimic plants. The sea dragon *Phycodurus eques,* A, mimics algae, while the leaf butterfly *Kallima paralakta* has wings shaped like a leaf. A, from Norman, *A History of Fishes,* 1963

103A), while a number of insects are remarkably similar to leaves (fig 103B).

Once again, behavioural adaptations have evolved which afford considerable protection to some animals. Most amazing is the shrimp-fish from the Indian Ocean. This curious creature swims in small shoals like any other fish but, if danger threatens, it can stand on its tail and continue swimming in the vertical position, for all the world like a piece of drifting weed (fig 104A). Another bizarre fish is the *Carapus* which escapes from its enemies by wriggling through the anus of a convenient sea cucumber or starfish! (fig 104B).

I have given, I hope, enough examples of adaptations in animals to stress the fact that every part of an animal is adapted to perform certain functions. Moreover, adaptations are extremely efficient for, as we have seen, an inefficient mechanism is an impossibility in the natural world because of the ruthlessness of natural selection.

This last fact poses certain problems to the biologist, and it is worth further discussion. It is easy to imagine why certain adaptations confer selective advantage on their possessors, but

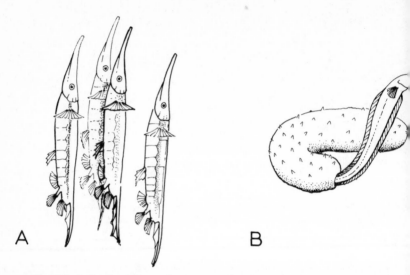

Fig 104 Animals with curious protective behaviour. Shrimp fish *Aeoliscus strigatus*, A, swim vertically to mimic drifting sea weed, while *Carapus*, B, hides in the anus of a sea cucumber. From Norman, *A History of Fishes*, 1963

it is far more difficult to see how they evolved in the first place. For example, both a limb and an ear are complex organs capable of performing important tasks. Structures like limbs, with their coordinated muscle, nerve, and bone systems, and the ear, with its ear drum, ossicles, and nervous supply, could not have evolved overnight, so what were these organs like when they first appeared? The theory of natural selection demands that organs are efficient at all stages of their evolutionary development; if they were not, their possessors would simply have become extinct. So how did the primitive limb and ear function successfully?

To begin with the limb. During Upper Devonian times, about 375 million years ago, there lived a fish called *Eusthenopteron* (fig 105A). It differs from the vast majority of fish of today in that its paired fins are not entirely composed of rays, but are supported by a fleshy lobe. The interesting fact about this lobe is that it contains a well-developed skeleton (fig 105B). The pectoral fin had two main functions: firstly, it could be flexed, or bent forwards, and so acted as a brake; and secondly, it could support the body while the fish was resting and may even have enabled *Eusthenopteron* to drag itself along the bottom of a shallow pond or river. *Eusthenop-*

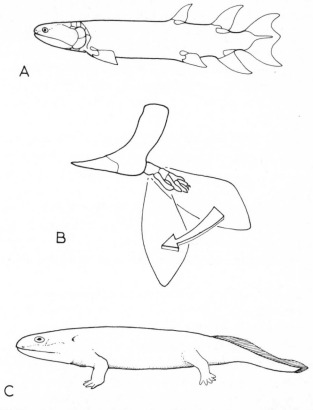

A

B

C

Fig 105 *Eusthenopteron*, a Devonian fish, A, and an enlarged view of its shoulder girdle showing the possible movements of its pectoral fin, B. Fish related to *Eusthenopteron* gave rise to the earliest amphibian, *Ichthyostega*, also from the Devonian, C. For explanation, see text. Not drawn to scale

teron is related to the lungfish, and it also probably had lungs, or at least, an air-bladder. This last fact has prompted some zoologists to suggest that *Eusthenopteron* was able to drag itself out of the water and use its pectoral fins to reach some new locality nearby. It is difficult to prove whether or not this was in fact possible. What is known, however, is that amphibians evolved from fish similar to *Eusthenopteron* and, by the end of the Devonian period, we find forms such as *Ichthyostega* that almost certainly could crawl overland (fig 105C). *Ichthyostega* has reasonably well-developed limbs and the main bones contained within them can be identified amongst those in the fin lobe of *Eusthenopteron*. In other words, the limb did not evolve as an entirely new structure; it represents a previously occurring organ that has been modified to perform a new function.

The ear in mammals is a complex organ and I do not intend to discuss the evolution of all its parts. Instead I wish to concentrate attention on the delicate mechanism composed of three articulating bones, called ossicles, that transmit vibrations from the ear drum to the oval window at the entrance to the inner ear (fig 106A). It is difficult at first to see how these bones, the malleus, incus, and stapes, could have been produced by the kind of random variation discussed in earlier chapters.

The story begins with *Probelesodon*, a reptile roughly the size of a large dog that lived in South America about 200 million years ago. At that time, mammals did not exist, and *Probelesodon* is a representative of the group of animals, the mammal-like reptiles, from which the mammals evolved. Surprisingly, before we can understand the origin of the mammalian ear ossicles, we must study the jaws of *Probelesodon*. Its lower jaw is composed of a large tooth-bearing bone called the dentary behind which there are a number of smaller bones (fig 106B). One of these small bones is the articular and, as its name suggests, it is this bone that articu-

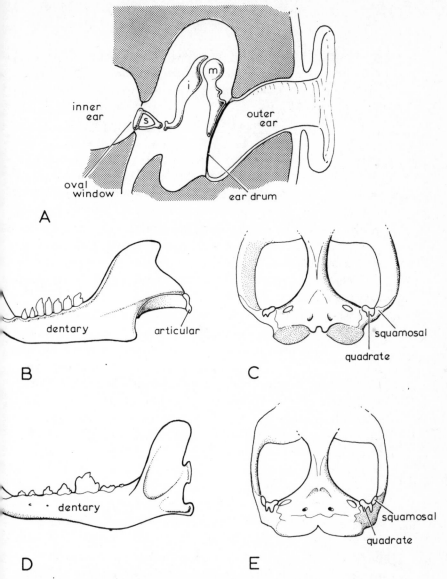

Fig 106 A section through the middle ear in man, A. The lower jaw of the mammal-like reptile *Probelesodon*, B, and the hind part of the skull of *Probelesodon* seen from below, C. The lower jaw of a mammal (a dog) D, and the hind part of the skull of *Probainognathus*, E. For explanation, see text. B and C drawn half natural size. From Romer, 'The Chañares (Argentina) Triassic Reptile fauna V', *Breviora*, No 333, 1969. E drawn half natural size. From Romer, 'The Chañares (Argentina) Triassic Reptile fauna VI', *Breviora*, No 344, 1970

lates with a part of the upper jaw, the quadrate. The quadrates are seen when the skull of *Probelesodon* is viewed from underneath (fig 106C), and each is socketed to a bone called the squamosal.

All reptiles have the kind of jaw articulation just described. A reptile can thus be partly defined as an animal with a jaw joint that occurs between the articular and quadrate bones. Another characteristic feature of reptiles is that a single bone, the stapes, transmits vibrations from the eardrum to the oval window; it will be remembered that this same bone is also present in the mammalian middle ear. One further point must be noted. One end of the stapes in living reptiles lies extremely close to the jaw joint and is sometimes actually attached to either the articular (in crocodiles), or to the quadrate (in young tortoises).

If we now look at the lower jaw of a mammal such as a dog (fig 106D) we find that it is composed of a single bone, the dentary, that articulates with the squamosal of the upper jaw. In other words, mammals have a jaw joint occurring between the dentary and squamosal bones, and the articular and quadrate of their reptilian ancestors appear to have been lost. A condition intermediate between that of reptiles and mammals is seen in another mammal-like reptile from the Triassic of South America. It is called *Probainognathus*, and its lower jaw articulates with a facet composed partly of the quadrate and partly of the squamosal bones (fig 106E).

During the evolution of the mammal-like reptiles, many changes occurred in the structure of the skull. Two that are important in this context are a tendency towards reduction of the height of the skull, and a change in the direction in which the muscle responsible for closing the jaws operated. As a result of these and other changes, the articular bone gradually lost its function and was replaced by a backward extension of the dentary. Eventually the dentary took over the function of the articular but instead of articulating with the quadrate, it formed a new joint with the squamosal. At some stage in the evolution of early mammals we must imagine an animal with a mammalian jaw articulation, and with two bones, the articular and quadrate, still present but no longer functional. Now, as I have noted earlier, the reptilian quadrate is often found in close proximity to the stapes, and it appears that, as they lost their primary function, the articular and quadrate bones became associated with the stapes and were transformed into ear ossicles. The problem of the origin of the mammalian ear ossicle is therefore essentially solved; the stapes corresponds to (or is homologous with) the stapes in reptiles, the incus to the quadrate, and the malleus to the articular.

I have dealt with the origins of the vertebrate limb and the mammalian ear ossicles at some length. They illustrate that adaptations, however complex, can be seen, not only as the products of selection of innumerable variations, but as organs and organ systems which have functioned effectively at every stage of their evolutionary history.

12

The Role of Animal
Behaviour

One of the clearest examples of the operation of natural selection has been already discussed in Chapter 9, and involved the change in colour of the moth *Biston betularia*. This change was caused primarily by a change in the environment (pollution), although the actual selection of individual moths was made by other animals, in this case the various birds that fed on *Biston*. But this is not the whole story. The researches that led to an understanding of colour change in *Biston*, also revealed that moths of different colours behaved in different ways. In an experiment, an equal number of dark and light moths were offered the 'choice' of

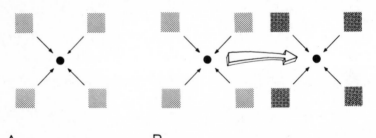

A   B

Fig 107 The effect of animal behaviour on evolution. In A, an animal (or population of animals) is affected by a number of environmental factors. In B, by changing its behaviour, an animal can bring itself into contact with a different set of environmental factors

light and dark backgrounds on which to rest. The results suggested very strongly that dark moths tended to settle on dark backgrounds and vice versa. The advantages of such behaviour are obvious, and show that we cannot afford to forget that animal behaviour plays a significant part in the evolutionary process.

By this, I do not mean to imply that an animal's 'desires' or 'wishes' play a part in its evolution; such talk is anthropomorphic nonsense. What I do mean, is that the evolution of many animals cannot be explained simply as the result of the effects of external factors alone. By changing its behaviour, an animal can effectively change the environment in which it lives, and can thus bring itself into contact with different environmental conditions that will favour different mutations. Figure 107 demonstrates what I mean. In the first case an animal, represented by a black dot, is subject to a variety of environmental conditions both physical and biological. If, now, this same animal behaves in a different way, moves to a new locality, feeds on a new food or what-

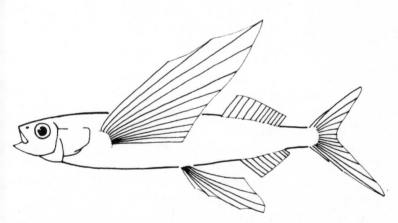

Fig 108 The flying fish *Exocoetus volitans*. It is unlikely that environmental factors alone can explain the evolution of its extreme adaptations. From Norman, *A History of Fishes*, 1963

ever, its environment is changed, and its evolution will be changed as a result.

The flying fish *Exocoetus* escapes from predators by leaping from the water and skimming its surface for considerable distances before entering the water again. It is able to do this because its pectoral fins are greatly enlarged and function as wings (fig 108). In water, such fins confer no great advantage and it is difficult to see how they could have evolved. If, however, we imagine that an ancestor of the flying fish with normal pectoral fins was in the habit of leaping from the water to escape predators we see that circumstances would encourage the selection of variants with large pectoral fins. In other words, habit or behaviour preceded the evolution of a definite morphological feature.

Before I continue, two very important points must be made. Firstly, mutations are random, and there is no evidence that behaviour can in any way encourage any particular mutation to take place. In the case of the flying fish, any mutation that tended to produce an enlarged pectoral fin would occur whether or not the fish was in the habit of leaping to escape predators. The point is this: before the fish developed the habit of leaping, mutations producing an enlarged pectoral fin would not confer any advantage, while the very same mutations would be of great advantage after the habit had been cultivated. The second point is that an animal cannot escape from the effects of natural selection by changing its behaviour; it can simply subject itself to different selection factors.

Natural selection acts on populations and not individuals. It follows that, before some aspect of behaviour can play a part in the evolutionary process, it must become a standard feature of at least a large part of any population. How does a behavioural trait spread throughout a population? There are two ways in which this is possible. Some kinds of behaviour are genetically determined; these are so-called innate or instinctive behavioural patterns, and we can imagine the spread of such behaviour by exactly the same means as are seen to operate in the case of the spread throughout a population of any physical characteristic that is genetically determined. But other kinds of behaviour are not genetically determined; many of the activities of man, for example, can be shown to be the direct result of environmental or cultural influences, and it is these kinds of behaviour that I want to discuss.

Once, it was thought that man alone could pass information from one individual to another, but this is not true as the following example shows.

For most of this century, milk has been delivered to householders in glass bottles sealed with cardboard or metal tops. Milk is, moreover, delivered during the early hours of the morning, so that for several hours each day an almost unlimited supply of food is to be found, free for the taking, on suburban doorsteps. In 1921, this fact was discovered by a tit, it is not known which species, in Swaythling near Stoneham, Southampton, and today it is common to see great tits, blue tits, coal tits, and other birds feeding from milk bottles. The behavioural patterns involved are complex. They involve recognising a milk bottle for what it is, and pecking through its top to get at the milk.

Records of observations of various tits feeding on milk have been made for the period 1930 to 1947 (fig 109). They show that the habit spread throughout the population so quickly that it is unlikely that we are dealing here with a genetically determined habit. The maps also show that the new behaviour was discovered to be profitable on many independent occasions, for it is known that tits rarely move outside a radius of some 15 miles. New behaviour, then, can spread throughout a population even though it is not genetically determined.

Unfortunately, not enough time has elapsed for us to know

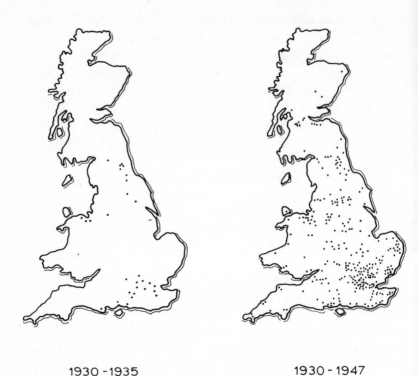

1930 -1935   1930 - 1947

Fig 109 Between 1930 and 1935 few examples of tits feeding from milk bottles were observed but, by 1947, the habit appears to have spread throughout most of the British Isles. From Fisher and Hinde, The opening of milk bottles by birds, *British Birds*, Vol 42, 1949

whether the new behaviour seen in tits will affect their evolution. The chances are that it will not, for it is unlikely that milk bottles will become a very permanent feature of the environment. Now let us look at another surprising phenomenon in birds, the feeding of the woodpecker finch, *Camarhynchus pallidus*, from the Galapagos Islands (fig 25, p 45). It may be remembered that this bird uses a cactus spine or small twig with which to probe for insects that live under the bark of trees. This habit is certainly genetically determined, and is one of the characteristics that distinguish the woodpecker finch from other finches of the Galapagos Islands. How could such behaviour have evolved?

The process of change that led to the evolution of the woodpecker finch is summarised in fig 110. To begin with, one or a few finches discovered that, by using a cactus spine, they could reach otherwise inaccessible supplies of food, A. This habit was copied by other birds and quickly spread, as did the habit of drinking from milk bottles in tits, until

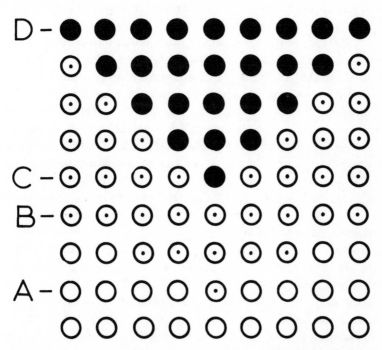

Fig 110 Change in a population stimulated by behaviour. Suppose that, at time A, a member of an animal population acquires a new habit. If this habit is beneficial it will spread rapidly throughout the population, B. Once the habit is established, any mutation that tends to facilitate it will also spread throughout the population, C–D

all members of the population had acquired the habit, B. At this point the woodpecker finch would have effectively changed its environment in that it would be feeding in a different way to that found in other finches. Within this new environment, any mutation that tended to facilitate the new habit, C, would confer selective advantage on its possessor, and would spread in the normal way until it became standard throughout the species, D.

The process just described has been called organic selection, and has only recently been accepted as part of evolutionary theory. For a long time examples of its operation were discredited because they were misinterpreted as the evolution of acquired characteristics.

The case of the woodpecker finch shows how a behavioural trait could possibly have evolved. Another example will serve to show that the same mechanism may well account for the evolution of physical characteristics. The most obvious difference between the skull of a dog and that of a cat is that in the former the snout is long and in the later it is short (fig 111 D and F). Figure 111 also shows that these have a common ancestry, and I include it simply to demonstrate why the dog and cat are so similar in most respects. Both the dog and cat are meat eaters and hunt live prey. The dog is a good runner and chases its prey over long distances. A highly developed sense of smell is important to the dog because it enables it to keep track of its prey during the initial stages of the hunt. For this reason the nasal organs are large, and are accommodated in the long snout (I am talking of the wild dog, not domestic varieties that often have atypical characters as a result of cultivation by man). A contrasted behaviour is seen in cats. The cat stalks its prey and makes its kill with a sudden pounce; eyes are well developed for this purpose and are therefore large, while the nasal organs and snout are small compared to those of a dog. Now, how do we explain these differences between cats and dogs? We

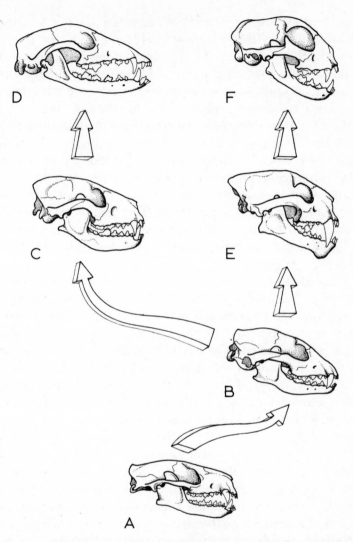

Fig 111 The evolution of cats and dogs from a common ancestor. A *Vulpavus* (Eocene), B *Pseudocynodictis* (Oligocene), C *Cynodesmus* (Miocene), D dog *Canis*, E *Dinictis* (Oligocene), F cat *Felis*. For explanation, see text. Not drawn to scale. A, B, C, and E from Romer, *Vertebrate Paleontology*, 1966: With permission of the University of Chicago Press

173

must, I think, assume that the different methods of hunting appeared first in the evolution of these animals, for it was only when the ancestors of dogs had adopted the habit of tracking their prey that mutations tending to enlarge the nasal organs and snout would have any selective advantage. The same argument applies to the ancestors of cats.

We have seen then, how the behaviour of an animal may affect its subsequent evolution. We may also note that habit must come before structure in the evolutionary sequence. Although some animals have habits for which they have no special adaptations, very rarely do animals possess physical characters that have no use. If the latter condition is found, it can always be shown to involve an organ that was useful to that animal's ancestors; such organs are called vestigial organs.

Although, as I have said, habit must come before structure, it is an oversimplification to regard the sort of evolutionary change under discussion as a process whereby a single behavioural trait is followed by the spread in a population of a single new mutation. The exact shape and size of the fin in the flying fish, and of the snout in carnivores, is determined by a large number of genes. In the same way, a behavioural trait is also affected by a whole suite of genes—not just one. We must therefore imagine the process summarised in fig 110 as consisting rather of a series of changes in behaviour, each creating conditions that favour the spread of one of a series of mutations that together produce an even more perfect adaptation. In other words, the beginnings of new behaviour may encourage the selection of mutations that, in turn, make possible further development of the behaviour.

Natural selection in a sense directed by behaviour, is probably the mechanism that produces the apparently directed change which is sometimes called orthogenesis. In Chapter 9, I mentioned the case of the Irish elk, *Megaloceros*. Fossils of this family form a series in which the antlers become increasingly large, until they attain an enormous size. These antlers were probably used during courtship battles, in which males with abnormally large antlers would succeed. Under these conditions, created by behavioural patterns, any mutation producing larger antlers would rapidly spread throughout the elk population.

13

The Origin of Species

Before discussing speciation or the origin of species, we must try to define the word 'species'. This is difficult because no precise definition can be made, but the effort is worthwhile because an appreciation of the reasons for our difficulties helps one to a better understanding of the evolutionary process.

We all know that a dog is different from a cat, and that each animal has characters that are unique. But a list of characters does not explain what a species is, it can only describe one particular species. We need a more general definition. The characters that are seen in a cat are determined by the genes that cats commonly possess, and variation amongst cats can be largely attributed to different combinations of these genes. All the genes that are found in cat populations may be considered as a unit, a gene pool. Now, the reason why cats are distinct from dogs is that the gene pool of cats is never mixed with that of dogs. It is not difficult to see why this is so. The bringing together of genes from different pools only happens when sexual reproduction occurs between animals from different gene pools. In the case of cats and dogs this doesn't happen, because cats never breed with dogs. Our definition of a species must therefore take into account the fact that species are distinct because they are isolated as far as reproduction is concerned.

A species, then, is a population of animals that breed amongst themselves but not with members of other popula-

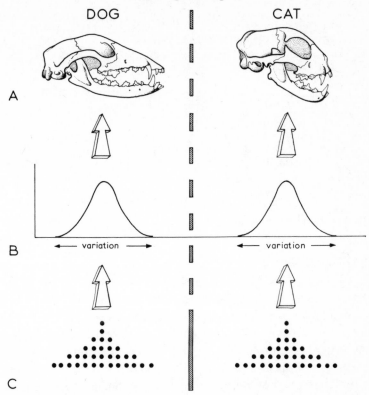

Fig 112 The definition of a species. The dog and cat are separate species. They have distinct characters, A, that are nevertheless subject to some variation, B. This variation is produced by genetic variation in their respective gene pools, C. There is a barrier that prevents mixing of genes from the pools of dogs and cats, thus the species remain distinct

tions (fig 112). Now let us discuss speciation in the light of this definition.

Simple change is change in a population over a period of time. It is therefore most directly seen by studying the fossil record (for example, taeniodonts fig 26, p47; and *Gryphaea* figs 42 and 44, pp73 and 75). Let us suppose that an animal passes through the stages represented by population samples

177

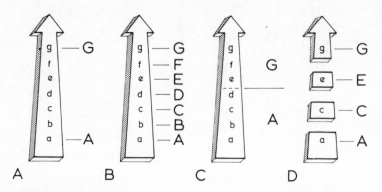

Fig 113 Different concepts of species in a changing population. For explanation, see text

a–g (fig 113A). It is clear that, because stage a is separated from stage g by a long period of time, members of those populations could never have bred with one another. They can therefore be assigned to separate species A and G according to the definition given above. We are able to define species A and G because, during the process of simple change, members of species G will have evolved physical characteristics that are not found in species A. So far there are no problems. Difficulties arise when we consider populations b–f. The characteristics of members of these populations are intermediate between those of species A and G, and are therefore difficult to define with precision. Do populations b–f represent separate species or not? In practice, the taxonomist may choose one of several alternatives: he may regard each population as representing a separate species (fig 113B); or he may assign some populations to species A and others to species G (fig 113C). If he takes the latter course, he must find at least one character that is distinctive and that serves to distinguish species A from G. Quite frequently, the taxonomist's problems are solved by the fact that complete series of fossils are rarely preserved. What he has is a sample, say of stages represented by populations a, c, e, and g. In this case the simple solution is to call each stage a separate species (fig 113D).

Enough has been said to show that the identification of an extinct species is almost impossible. Firstly, the members of an extinct species can only be defined by their physical attributes, not by their ability to breed with one another. Secondly, even a description of a fossil animal's physical appearance is incomplete, for it must necessarily include description of only those parts that are preserved. Lastly, the divisions between species may be defined by chance gaps in the geological succession or by changes in arbitrarily chosen morphological characters. These same limitations apply to fossil forms that are subject to divergent change.

Having appreciated some of the difficulties encountered

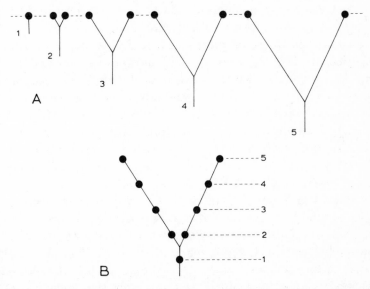

Fig 114 A large number of animal populations show different stages of divergence A 1-5. The stages 1-5 can be combined to produce a model for the evolution of two species by divergent change B

when dealing with extinct animals, one might suppose that to identify a living species is a far simpler task. In fact, this is not so. Divergent change is taking place all the time and, at any particular period of time, there are large numbers of animal populations that are at different stages of divergence. In fig 114A, stage 1 represents a single species and stage 5, two species that have evolved from a common ancestor. In each case, the species is represented by a population, the members of which breed amongst themselves but not with members of other populations. But the change from one species to two descendant species is not an abrupt one, it is gradual. Many populations occur, such as those represented by stages 2–4, in which the individuals of each population have an increasing tendency to keep themselves to themselves, but which are not completely isolated from one another. Let me give some examples.

An example of populations that have diverged but little from a common ancestor (stage 2) are those of a moth, *Hyponomeuta padella*. This species can be divided into two subspecies. In one, the eggs are laid on apple leaves and the adults are usually dark grey, while in the other, the eggs are laid on hawthorn leaves and the adults tend to be lighter in colour. Members of these subspecies occasionally mate with one another, although breeding experiments show that about twice as many matings take place between individuals from the same subspecies as take place between individuals from different subspecies.

Stage 3 may be represented by two species of the fruit fly *Drosophila*, *D. pseudoobscura* and *D. persimilis*. These species are almost identical but have somewhat different geographical distributions, the latter being limited to a relatively small area on the west coast of the United States, while the former is more widely distributed. It is unlikely that members of these species normally mate with one another, although mating has been observed in the laboratory. When mating does take place, the resulting male offspring are infertile, but the close relationship between the two species is demonstrated by the fact that the females produced by cross matings are fertile.

Finally, the horse and the ass provide us with an example of stage 4. These species, *Equus caballus* and *E. asinus*, have quite distinct morphological characters and never breed with one another in the wild. However, they have not sufficiently diverged from their common ancestors to prevent mating in captivity, although their offspring, called mules, are almost always sterile.

These few examples show that speciation is still going on, and that populations can be found that represent every stage of divergence. The different stages of divergence seen in animal populations of today can be combined to produce a model for the stages through which any two species with a common ancestor must have passed. This is done by arranging the stages 1–5 with a vertical time scale (fig 114B).

In conclusion then, we may appreciate the process of speciation by studying, firstly, closely related living populations, their morphological differences, and their ability to interbreed and, secondly, the history of the divergence of animal populations as evidenced by the fossil record. In the latter case, we can only use morphological criteria as evidence of divergence because it is not possible to know of the breeding habits of animals that are now extinct.

A more detailed model of divergence is shown in fig 115 in which the dots represent different genes, and in which species are represented as isolated gene pools. Subspecies (or very closely related species) are represented by gene pools that are not completely isolated because some interbreeding is possible.

We must now consider the forces that cause divergence in animal populations. This must be done in two stages; first we must identify the forces that cause fragmentation, then

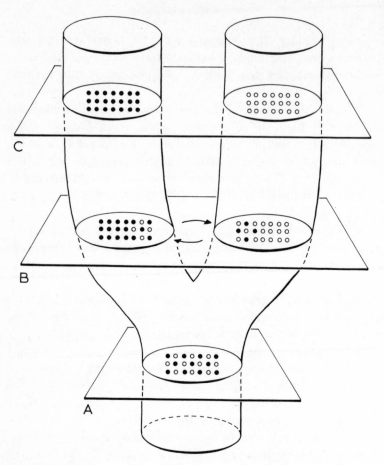

Fig 115 A model of species divergence. The diagram may be inter-
preted in two ways. Firstly, the dots may be regarded as genes of a
gene pool. At time A, a gene pool representing a single species occurs.
At time B, the gene pool has become partially divided, each part has
distinctive features although some mixing occurs. These represent subspecies. At time C, the two gene pools have become completely isolated and two new species can be said to have evolved from the one
present at time A. A second interpretation of the diagram is possible.
The dots may be seen as representing individuals (or groups of individuals) forming discrete populations in areas symbolised by the
ellipses. At time A, a single population occurs in one geographical
area. At time B the population has spread to a larger area, and at time
C two separate areas are colonised

182

we must consider why populations, having once become fragmented, remain isolated from one another.

Current opinion amongst biologists is that fragmentation of a gene pool is, in most cases, the result of the spread of a species over a large or varied geographical area. To explain this, we may turn again to fig 115 and reinterpret it as follows. The ellipses may now be seen as geographical areas and the dots as individuals or groups of individuals in a population. At a moment in time, A, a species is limited in its distribution to one particular area. If this species spreads, we may imagine a situation at time B, in which a new enlarged area is occupied by that species. Let us suppose that this new area is characterised by slightly different ecological conditions that correspond to the areas indicated by the two ellipses. In each area, natural selection will have different effects on the population so that after a time, divergence will take place and subspecies will be produced. Initially there will be a certain amount of cross breeding between members of these subspecies, but eventually the two populations will become distinct (fig 115C). At this point, the process of fragmentation is complete, but the resulting populations will only evolve into separate species if they are isolated from one another in the sense that interbreeding becomes impossible. A number of isolating mechanisms have been discovered and these will be discussed later in this chapter.

I wish first, to give some examples of populations that have spread over large areas and in which fragmentation has begun. Such populations can be broadly divided into three groups: those in which certain individual characteristics vary from one region to another (clinal variation); those in which parts of a population are geographically separated from one another, or are found on the edge of the area occupied by the main part of the population: and those in which two populations are separated by a narrow zone in which hybrids are common.

183

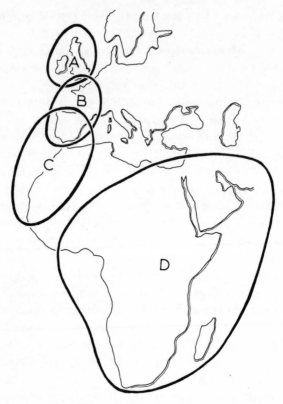

Fig 116 The winter ranges of four populations A–D of the plover *Charadrius hiaticula*. Based on a drawing from Salomonsen, 'The evolutionary significance of bird-migration', *Ogn. Biol. Medd.*, Vol 22, 1955

Examples of the first group are seen in plover and herring populations. There are four populations of the plover *Charadrius hiaticula* that spend winter in different parts of Europe and Africa (fig 116). The mean wing length of adults in each of these populations varies from 134.9cm in the northern population to 125.3 in the southern one. This variation is likely to be associated with different climatic conditions. In the Pacific herring *Clupea pallasi*, the number of vertebrae in the backbone varies from twenty-two to twenty-four as one passes along the coast of North America from California to Alaska (fig 117). Again, this variation is probably correlated with water temperature differences in different parts of the area occupied by the herring.

An example of a population that has spread over a large area and which is now represented by a group of isolated species is seen in the tree creepers of Australia (fig 118). These birds vary in colour from almost entirely black to light brown, and occupy savannah woodland areas. With one exception,

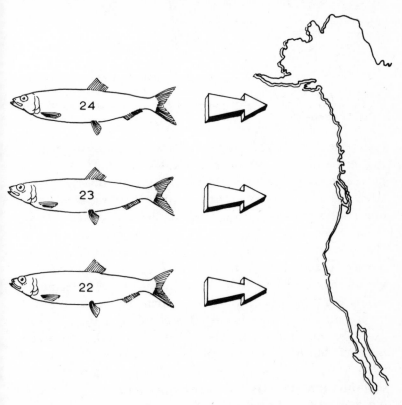

Fig 117 The number of vertebrae in the backbone of Pacific herrings, *Clupea pallasi*, occurring along the western coast of North America

185

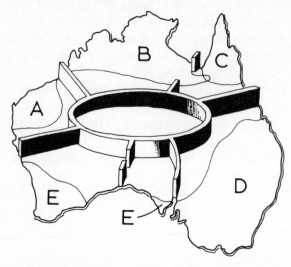

Fig 118 Barriers separating populations A–E of tree creepers in Australia. For explanation, see text

each population is separated from its neighbour by dry regions that are unsuitable for these birds and which act as barriers preventing cross-breeding. As a result, greater divergence has taken place amongst these tree creepers than was seen in the first two examples, and each population is considered to be a separate species.

A variation of the type of population exemplified by the tree creepers is seen in the fly *Phlebotomus papatasii*. This species has a distribution throughout the Mediterranean region (fig 119) and, as there is a reasonable chance of cross-breeding between members of this population, there is little variation amongst its members. Divergent forms do occur, however, but only at the extreme edge of the area occupied by this fly. Their occurrence can be explained by the fact that their position makes cross-breeding with the main breeding stock rather unlikely.

The last kind of population mentioned above is well illus-

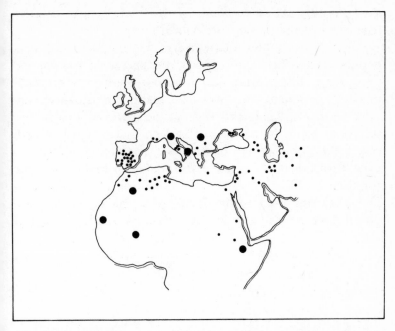

Fig 119 The fly *Phlebotomus papatasii* is distributed throughout the Mediterranean region. Divergent forms (large circles) almost invariably occur at the edge of their area distribution. From Hennig, *Phylogenetic Systematics*, 1966. With permission of the University of Illinois Press

trated by the European crows. The carrion crow and hooded crow represent populations that have only recently diverged from a common ancestor and, although each kind is found in a different part of Europe, there is a hybrid zone between the two areas (fig 120).

Of course, not all animal populations fall into the simple classification given above. *Cerion* is a land snail found along the east coast of Cuba. It is limited to a strip of vegetation near the high tide mark. Along a thirty mile stretch of coast, there are points that are unsuitable for this snail, and these separate populations that have become morphologically distinct (fig 121). At other points, perhaps where such barriers have been removed, there are hybrid zones.

In all the examples mentioned so far, fragmentation of a population has been preceded by the spread of the population over a wide geographical area. But there are other ways in which fragmentation may take place. For instance, consider an insect that lays its eggs on the leaves of a particular species of plant. Its larvae will feed on that same plant and, unless there is a period of dispersal, the young adults will mate with other adults that have grown up under the same conditions. Now, if by chance a female lays her eggs on a different plant, the resulting larvae will grow under new conditions and natural selection may favour variants that would

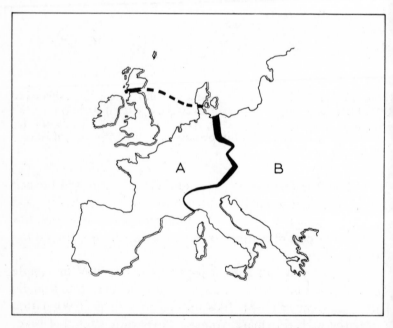

Fig 120 The hybrid zone between populations of the carrion crow A and the hooded crow B. Based on a drawing from Meise, 'Die Verbreitung der Aaskrähe', *Journal of Ornithology*, Vol 76, 1928

Page 189 Protective colouration in fish and birds. The colour of the bullhead, *Cottus gobio (above)*, serves to make its outline less conspicuous, while that of the black grouse, *Lyrurus tetrix (below)*, makes it almost invisible as it sits incubating its eggs

Page 190 (above) There are two ways in which changes take place in animal populations. Genetic change is the result of differing rates of reproduction of genetically different individuals, while cultural change is the result of the spread of new types of behaviour by imitation. Blue tits commonly feed on milk from bottles. This behaviour is not genetically determined, which explains why it spread throughout the tit population in Britain so rapidly; *(below)* drawings made in the field of the woodpecker finch, *Camarhynchus pallidus,* using a cactus spine as a tool with which to probe for insects. This behaviour is genetically determined, but must have originally spread throughout the finch population by imitation

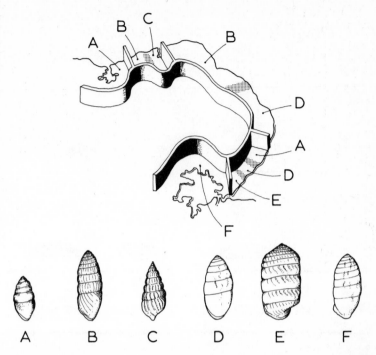

Fig 121 Barriers separating populations of the snail *Cerion*. Hybrid zones marked in stipple

not normally have survived. If these variants breed together and in turn lay their eggs on the 'new' plant, the genes determining their peculiar characteristics will be preserved instead of being mixed into the gene pool of the parent species. Gradually a population will become defined that is significantly different to the parent population, and a new species may arise.

Such may have been the history of some closely related species of sawflies that lay their eggs on different species of willow trees, and of the subspecies of *Hyponomeuta* mentioned earlier in this chapter.

Fragmentation does not always result in the production of

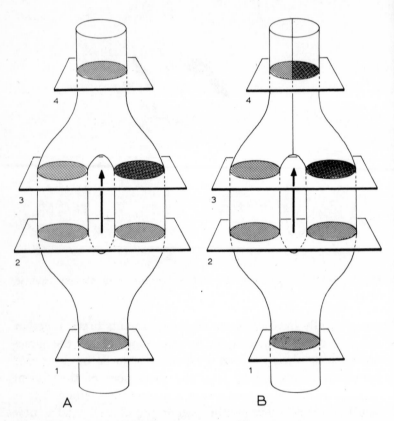

Fig 122 A model of species divergence. In A, a species at time 1, forms separate populations by time 2. During the period marked by the arrow, divergence takes place, so that the populations are distinct by time 3. When the two populations come together, mixing of the gene pools results in the re-establishment of a single species by time 4. In B, an isolating mechanism is established that prevents mixing of gene pools after time 3 so that the populations preserve their identity and become new species

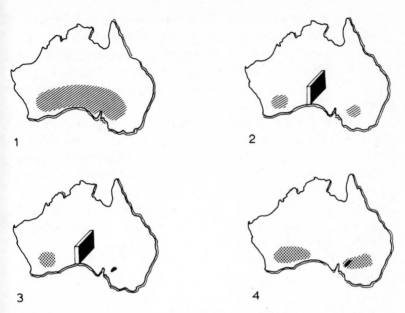

Fig 123 The history of the divergence of a population of Australian mallee thickheads follows the pattern illustrated in fig 122B. Numbers correspond to stages in fig 122B

new species. Two populations separated by a barrier may well come together again if that barrier is removed. This may happen even if during the intervening period the two populations have diverged (fig 122A). Speciation will only occur if some character evolves in one or both the diverging populations, that makes cross-breeding impossible when the barrier is removed and the populations are allowed to mix freely again (fig 122B). Such a character is called an isolating mechanism. Figure 123 suggests the history of two species of Australian mallee thickheads in which the stages 1–4 correspond to the stages 1–4 in fig 122B. At stage 2 a population has become divided by a barrier, in this case a newly developed region of aridity. Divergence takes place (stage 3), and we must suppose that amongst the newly evolved characteristics of the two populations, there are some that act as isolating mechanisms. When by stage 4, the barrier has disappeared and one of the divergent populations has spread to the east, no cross-breeding takes place. The divergent populations preserve their identity and are therefore to be considered new species.

A large number of isolating mechanisms have been identified in the animal kingdom. I will not enumerate them here; it suffices to note that they can be classified into three groups: those that operate after copulation so that either fertilisation or development of the embryo is not successful; those that prevent potential mates from copulating; and those that prevent potential mates from meeting at all. An isolating mechanism of the first kind operates between two frogs, the bronze frog (*Rana clamitans*) and the bullfrog (*R. cates-*

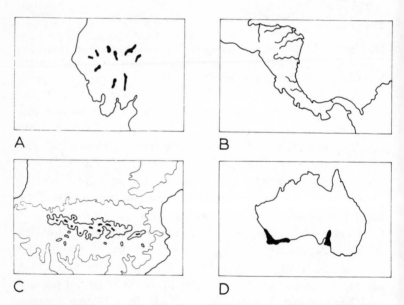

A

B

C

D

Fig 124 Ecological barriers preventing breeding between populations. A isolated lakes, B isolated river systems, C isolated high ground, D areas separated by deserts

beiana). Both occur in the western half of the United States but, if they mate, development of the offspring always fails. Earlier in this chapter I quoted a similar but less extreme example, the horse and ass are separate species because their offspring, mules, are infertile.

Isolating mechanisms preventing copulation are numerous and involve differences in the sex organs, and differences of body size and behaviour. The latter kind are of particular interest in the case of birds in which calls made during the spring by males are necessary to stimulate the females. There are closely related species such as some warblers, in which the song of the males is the most distinguishing feature.

Finally, there are a large number of isolating mechanisms that prevent members of two populations from meeting. These often take the form of geographical or ecological barriers and may be the same phenomena that caused initial fragmentation of a particular population. A variety of such mechanisms is illustrated in fig 124.

14

Broad Patterns of Evolution

If a zoologist were to study the workings of every kind of cell in the human body, he would learn a great deal. But parts of man would still remain a mystery, for many aspects of life are only apparent if the whole body is studied as a unit. Yet more can be discovered if groups or societies of men are investigated. In the same way, some aspects of the evolutionary process can only be appreciated if whole groups of animals are studied. Before discussing these broad patterns of evolution, let me summarise what we have learnt so far. In this way we may relate the subject of this chapter to preceding ones.

We have seen that species are interbreeding populations of individuals: looked at another way, a species is an isolated gene pool. Differences in the rate at which different members of a population breed, lead to changes in the overall constitution of the gene pool, thus, simple change takes place. As a result of the division of populations by what are often accidents of geographical and ecological distribution, gene pools are sometimes fragmented, and, in the various fragments different changes may take place. After a time, isolating mechanisms evolve, preventing subsequent mixing of the fragmented gene pools. When this takes place new species are said to have evolved by divergent change.

Because change takes place as a result of differential breeding, it always occurs so that individuals most suited to prevailing conditions constitute a majority in any population. This must be so, because by definition, individuals most suited to prevailing conditions are those which produce most offspring.

Over long periods of time, repeated divergence has led to the evolution of a vast number of species. In any particular group, most or even all the species may now be extinct. The relationships between the members of a group may be expressed in the form of a phylogenetic system. (A more everyday name for such a system is a family tree, but this name is misleading because, of course, the group of animals under discussion may not formally constitute a family.) A

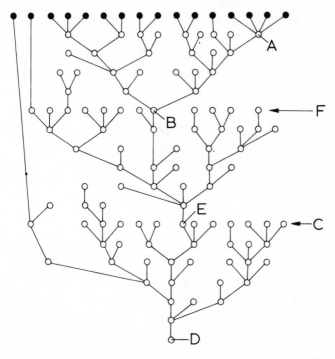

Fig 125 A hypothetical phylogenetic system or 'family tree'. White circles represent extinct forms, black circles, living forms. Lines represent phylogenetic relationship. For explanation, see text

hypothetical phylogenetic system is shown in fig 125, and some of its features will serve to illustrate the more important broad evolutionary patterns.

To begin with adaptive radiation. An adaptive radiation is the phylogenetic pattern that symbolises the divergence of a number of species from a common ancestor. On a small scale, the four living species which have evolved from species A constitute an adaptive radiation, and an example that we have seen before is the radiation of Darwin's finches. On a larger scale, we see the radiation of fourteen living species, and many now extinct, from a common ancestor B. A large scale radiation such as this is exemplified by the mammals (fig 126). The earliest known mammals occur during the Triassic. They were small mouse-like animals that fed on insects, and only two or three such families are known. During Cretaceous times there appears to have been little divergence for, by the end of that period, no more than ten families are known to

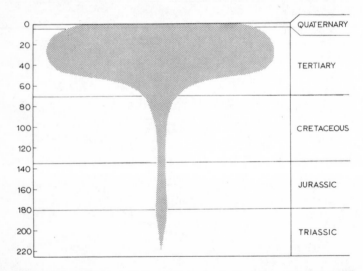

Fig 126 The radiation of the mammals during the past 220 million years. The width of the stippled area is proportional to the number of families known at any given time

198

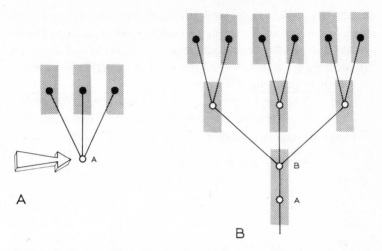

Fig 127 Radiation of species (black circles) from extinct species (white circles). In A, an immigrant species A gives rise to three new species that colonise new environments (stipple). In B, a species A evolves some adaptation B that makes an adaptive radiation possible

have evolved. Then, suddenly, about 60 million years ago during the Tertiary, we find the remains of animals that have been classified into some fifty families. At this time, well-known groups such as the rodents, carnivores, horses, and pigs have their origins.

When and why do adaptive radiations occur? Some radiations occur when an animal is introduced to a new area; a finch to the Galapagos Islands, or a marsupial mammal to Australia. When this happens, the immigrant species comes into contact with a large number of new environments, with the result that, if its gene pool becomes fragmented, the different fragments are likely to become adapted to a variety of conditions (fig 127A). Another cause of radiation is the evolution of a particularly successful adaptation in a species. By a successful adaptation, I do not mean an efficient means of obtaining one particular food or of surviving under a particular climatic condition, but rather an adaptation that enables its possessor to feed efficiently on a variety of food-stuffs or to live in a variety of climates. Descendants from a species with such general advantages are likely to be successful and will, in turn, tend to produce further species by divergent change (fig 127B).

If we look at the history of the vertebrates (fig 128), we find that each major group has passed through at least one phase of rapid divergence, while one group, the bony fish, has

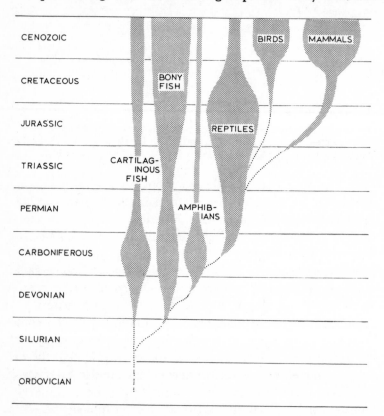

Fig 128 The fossil record of the vertebrates. From Romer, *Vertebrate Paleontology*, 1966. With permission of the University of Chicago Press

passed through two. In each case, the story is the same. In each case, some general adaptation evolves that gives the members of a group a distinct advantage over their competitors. In the case of the earliest bony fish, it was the evolution of an efficient locomotive system of fins and tail; a later radiation began with the evolution of improved methods of swimming and of a more flexible feeding mechanism. The amphibians owe their early success to the evolution of a limb that made overland travel possible, and the reptiles to features such as the aminote egg that made life independent of the water possible for the first time. The mammals owe their success to two adaptations; they are able to control their blood temperature, and most retain their young until a late stage of development (fig 20, p40). Because it is unlikely that these features have evolved separately in each mammalian family, it is reasonable to assume that they were present in Triassic and early Cretaceous forms. Now, both these features confer great advantages on their possessors; a constant blood temperature provides a stable internal environment that allows activity which is independent of changing conditions outside, and the retention of young inside the mother means that off-spring are protected during the most vulnerable part of their lives. Once these adaptations had evolved, it was not surprising that mammalian populations spread and diverged to colonise a vast range of habitats. Their adaptive radiation was assured.

The second broad evolutionary pattern, faunal replacement, is really a consequence of adaptive radiation. Returning to fig 125, let us consider the species living during the period of time C. These constitute a radiation from the species D, but one of their number, E, obviously had some special advantage, for it has given rise to a new radiation that culminates in the species living during the period of time F. It is quite likely that the descendants of species E occupy the same habitats at period F as were occupied by descendants of species D at period C. In other words, the old fauna has been replaced by a new one. (For an actual example of such faunal replacement during the history of the bony fish, see fig 27, p40.)

In fig 125, the old fauna became extinct before the radiation from species E began. This sometimes occurs, but on other occasions the replacement of one fauna by another is more gradual. The multituberculates were primitive mammals, known mainly by their curious teeth. During the Middle Palaeocene they reached their zenith, but they were then gradually replaced by similar mammals, the rodents, which diverged rather suddenly during the Lower Eocene (fig 129).

The examples quoted above are ones in which closely

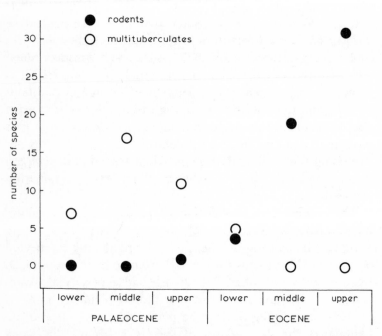

Fig 129 The replacement of multituberculates by rodents during the Palaeocene and Eocene

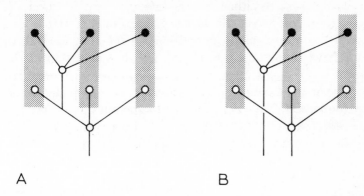

Fig 130 The replacement of faunas. In A, closely related groups replace one another. In B, a group replaces another that is distantly related

related groups replace one another (fig 130A). Sometimes a group can be replaced by another that is only distantly related (fig 130B), an obvious example being the replacement of flying reptiles, the pterodactyls, by birds.

The last major evolutionary feature to be discussed here is, appropriately enough, extinction. Extinction is the inevitable fate of a species just as death is of an individual. In fig 125, sixteen species are living while over a hundred are extinct; this is no exaggeration of what is true of many recent groups. Extinction may be due to three causes: intraspecific competition, interspecific competition, and what may be called external factors. Intraspecific competition is just another name for differential breeding within a population, which as we know, produces change in a species gene pool. When species A changes to produce species B, A is by definition extinct as soon as the change is complete. Interspecific competition is competition between different species. We saw two examples in fig 130. This figure does not, however, make clear whether later species replaced earlier ones, or whether they simply came to occupy habitats already vacated by earlier ones. In either case the fate of the earlier species is extinction, but it is only in the former case that it is due directly to interspecific competition. In the later case, there must have been some other cause for the extinction of the earlier species, such causes may be called external factors. We have seen an example of an external factor, a change in environmental conditions, causing extinction of the amphibian *Diplocaulus* (fig 69, p149), and it is apparent that there is no real difference between that example and fig 130A.

15

Man

There appear to be two major misconceptions concerning the origin of man. The first is that he is fundamentally different from all other animals and therefore has unique origins, and the second is that he is descended from an ape. Both ideas are wrong, but both contain a germ of truth.

At the end of the seventeenth century, Edward Tyson dissected a chimpanzee and realised that its anatomy was remarkably similar to that of man. As a result of work by

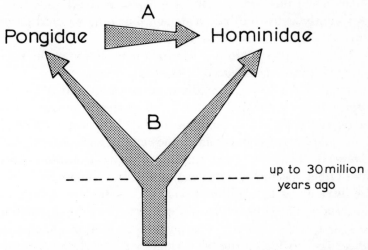

Fig 131 Man (Hominidae) did not evolve from the chimpanzee (Pongidae) A, but from a common ancestor B that lived up to 30 million years ago

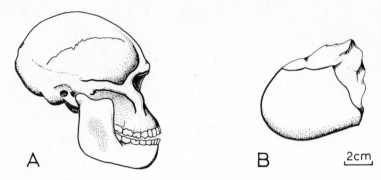

Fig 132 The skull of *Australopithecus africanus*, A, and an example of the simple tools he was probably able to make, B

later anatomists, the family to which man belongs (Hominidae) and that containing the chimpanzee (Pongidae), are classified in the same superfamily. This classification does not imply that man has evolved from the chimpanzee, but that both animals have a common ancestor (fig 131). We do not know when this common ancestor lived, it could have been as recently as five million years ago, but others would guess that 30 million is a more likely figure.

In this chapter I will first outline the evolution of the family Hominidae as evidenced by the fossil record.

We must begin with remains that have been called *Australopithecus*, a genus that contains two species, *A. robustus* and *A. africanus* (fig 132A). Some of the best-known specimens of *Australopithecus* come from the lowermost rocks in Olduvai Gorge in Tanzania, rocks that are about two million years old. Other specimens have been found in Ethiopia that may be three and a half million years old, but the age of these is uncertain. The bones of *Australopithecus* have characters that are comparable to those of apes and man. On the one hand, the lower jaw is massive, the brow ridges well developed and the brain cavity small; on the other, the teeth are similar to human teeth and the gait upright or almost upright.

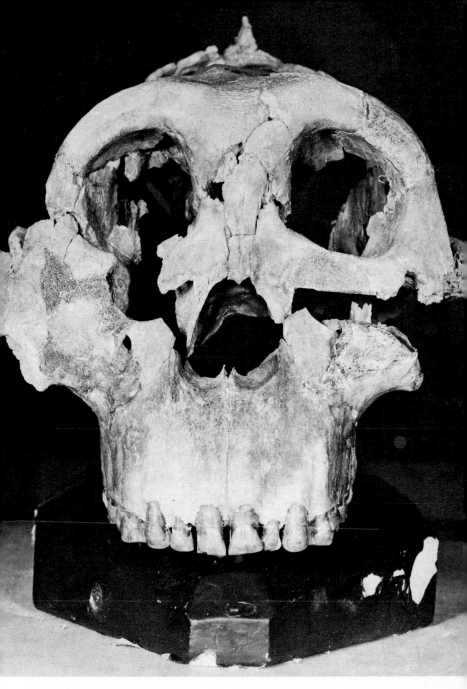

Page 207 A skull of *Australopithecus* from the Olduvai Gorge in Tanzania. Note the pronounced brow ridges

Page 208 Rock is carefully removed with dental picks from the fossilised remains of a palate. This picture was taken at Olduvai Gorge, Tanzania, the site of so many discoveries that have helped to elucidate the story of the early evolution of man

Later fossil men are classified in a single genus, *Homo*, and there are two species, *H. erectus* and *H. sapiens*. *H. sapiens* is subdivided into three subspecies, *H. sapiens steinheimensis*, *H. sapiens neanderthalensis*, and modern man, *H. sapiens sapiens*. In the past, a number of other names have been given to remains of fossil man (fig 133), but it is now generally agreed that these represent different populations and do not warrant formal identification in the classification. The reason for this is that the differences between forms such as Pekin man and Java man, are no greater than might be expected in a random sample of remains of modern man.

The history of the two species of *Homo* is one of repeated divergence of a large number of populations during some 600,000 years. This is indicated diagrammatically in fig 133. As can be seen, only a small sample from these populations is known, but they are enough to indicate that certain evolutionary trends have taken place.

Homo erectus is represented by fossils from Java, China, East Africa, North Africa, and Germany. Its skull is thick, its brow ridges and lower jaw heavily constructed, and its forehead low. The brain cavity is small compared to that of modern man, but larger than that of *Australopithecus*. Fossils of *Homo sapiens* have been found in Europe, the eastern end of the Mediterranean, Java, and in Africa. The three subspecies show a gradual lightening of the skull bones and an increase in the size of the brain cavity.

If we compare the series of skulls in fig 133 with similar fossil series (eg, figs 26 and 111, pp43 and 173), little obvious change appears to have taken place. This is partly because the changes illustrated in fig 133 are changes that have occurred over a period of only 600,000 years. Nevertheless, this emphasises the important point about man's evolution: significant changes were not necessarily of the kind that are readily preserved. Modern man is different from other animals, for example, in that he is able to communicate in a

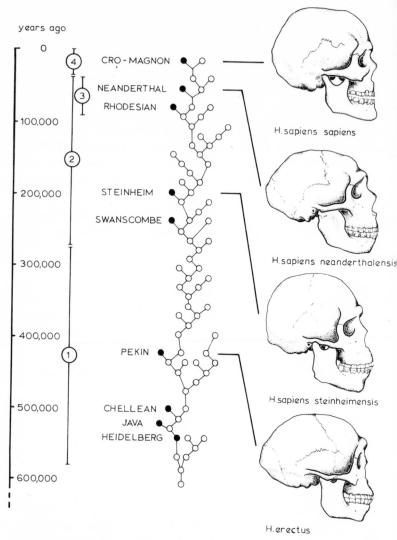

Fig 133 The history of man is one of repeated divergence of a large number of populations. Known examples are marked as black circles on the phylogenetic system. The ranges of the various species and subspecies of man are given on the left. 1 *Homo erectus*, 2 *Homo sapiens steinheimensis*, 3 *Homo sapiens neanderthalensis*, 4 *Homo sapiens sapiens*

variety of ways with his fellows, and the development of such abilities only rarely leaves traces in the fossil record. There are some, however. To begin with, remains of *Australopithecus* have been found in caves that also contain piles of bones of giraffes, hippopotamuses, and other smaller mammals and birds. The surprising fact about these remains is that most have fractured skulls and appear to have been killed with a blunt heavy weapon. *Australopithecus* then, had acquired a new character, the ability to use weapons or tools. The first tools were probably large stones, but more sophisticated tools must have been used as well. They were lumps of hard rock, usually lava, that had been roughly chipped so that at least one edge was sharp and could be used for cutting (fig 132B).

Specimens of *Homo erectus* from China (Pekin man) have been found associated with charcoal and ash, the earliest evidence of the use of fire. Tools have also been found (fig 134A) that indicate that *H. erectus* had learnt to chip flint with a fair degree of precision. Later deposits that are con-

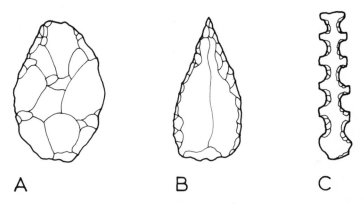

A   B   C

Fig 134 The gradual sophistication of flint tools during the history of man may be inferred from these examples. They are associated with *Homo erectus*, A, *Homo sapiens steinheimensis*, B, and *Homo sapiens sapiens*, C

temporaneous with *H. sapiens steinheimensis* contain tools that show an even greater degree of sophistication (fig 134B), while, by the time *H. sapiens sapiens* had evolved, flints that must have had specialised uses occur (fig 134C). It is likely that the flints just mentioned were made by specialist craftsmen and, if this is so, we have evidence that division of labour had developed in man's society. This in turn implies that communication between individuals was very likely possible.

As I pointed out at the beginning of this chapter, there are surprisingly few major anatomical differences between man and related animals. His one really important characteristic is the size and organisation of his brain.

Australopithecus had a brain cavity of 440 to 590ml, whereas in Cro-magnon man, *Homo sapiens sapiens*, it was about 1,500ml. This remarkable increase in brain size took place in less than two million years, and can be paralleled by the advances in technology that took place over the same period of time. This is no accident, as the two phenomena are certainly closely related. The evolution of the brain made the development of tools possible, and the development of tools made further enlargement of the brain of selective advantage. I have discussed similar examples of evolutionary change associated with behaviour in Chapter 12.

In the case of man, let us assume that *Australopithecus* had learnt to manipulate simple tools. We know that this is a reasonable assumption because of the associated finds mentioned earlier. Occasionally, mutations would occur in populations of *Australopithecus* that would produce individuals with slightly larger or, in some sense, more efficient brains. These individuals would be most likely to invent new tools and, when they did so, the social group to which they belonged would benefit, because the invention of a new tool implies greater control over the environment. By this, I mean that a new tool could make hunting more efficient, or make new foods or materials available for exploitation. Now, three things would happen to a group or population with a more advanced technology than its neighbours: firstly, it would be better equipped to survive periods of hardship; secondly, its members would tend to multiply and spread more rapidly; and lastly, its members would become dependent upon the new technology. Thus, populations with inventive members would gradually replace less endowed groups. Also, because successful groups would become dependent on the new tools, there would be selection within the population for individuals who were capable, not only of using tools efficiently, but of refining them still further. Such individuals would be those carrying mutations or gene combinations for large brains. In this way, continuous selection for increasingly larger brain size has generated technological development. The process is cyclic (fig 135), and explains why both phenomena have developed so rapidly.

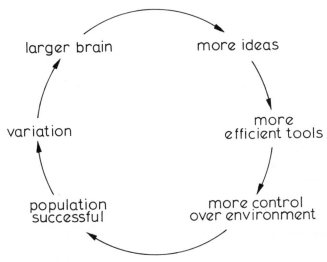

Fig 135 The relationship between brain size and the development of more efficient tools during the history of man. For explanation, see text

Man

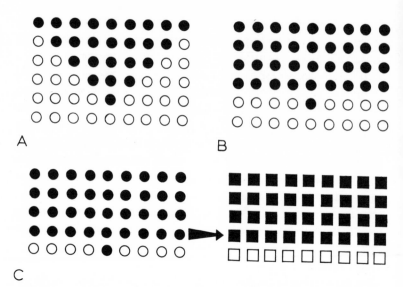

Fig 136 Differences between the spread of genetic and cultural information. A beneficial gene mutation can only spread throughout a population after a number of generations, A. A beneficial idea on the other hand, can be passed to all members of a population in a very short space of time, B. It can even be passed from one population to another, C

The evolution and spread of methods of communication can be explained in a similar way for, although the beginnings of language are shrouded in mystery, it is not difficult to see how any advance in communications would have had immediate and beneficial effects on its inventors.

This leads me to the next important fact about the evolution of man. Once communication became possible, an entirely new type of evolution occurred in human populations, I mean the evolution of ideas. There is no doubt that man's success is due mainly to the fact that he is able to organise himself into communities, to share common tasks, and to take on specialised tasks for the good of the community. All these things depend on an efficient means of communication and on the propagation of ideas. Development of these things during the history of man has been extremely rapid, and this is because an idea is entirely different from a mutation. If a beneficial mutation occurs in an individual in a population, it can only be passed to that individual's offspring. The mutation becomes established, only after many generations, when such offspring have replaced all, or most, other members of the population (fig 136A). A good idea on the other hand, can spread far more rapidly. With efficient methods of communication, an idea generated by an individual in one generation, can be the property of all members of subsequent generations (fig 136B). Also, an idea can be carried by a single individual from one population to another (fig 136C). Obviously, the propagation of ideas is never instantaneous, but it is far more rapid than the spread of a mutation can ever be.

Many people think that man is no longer evolving. This idea is false for the following reasons. We have seen that for an animal population to change a number of conditions must be satisfied. Firstly, variation must be generated by new mutations and gene combinations and, secondly, some variants must be more advantageous than others and contribute more offspring to successive generations. If these conditions are satisfied, simple change will occur. In addition, divergence will take place if parts of the genetic pool of a species become isolated and subjected to different environmental factors.

Do these conditions apply to man?

There is little doubt that, with respect to the first condition, man is not in any way significantly different from any other animal. We know of a large number of characters that owe their variability to varied genetic control, and an even greater number that are determined, partly by genetic, and partly by environmental factors.

One can be less certain whether many human variants are

215

poorly suited to their environment. The point about man is that he is able to exert fantastic control over his environment and can adapt conditions to suit himself. Obviously a few individuals inherit genes that cause such things as haemophilia, and these people have little chance of survival under any conditions, but most variants in countries that have reasonably high standards of medical knowledge can, and do, survive. It appears therefore, that natural selection has little effect on human populations today. But can we therefore conclude that no sector of the human race is contributing more individuals to successive generations than are others? The answer is no. Mainly as a result of advances in applied medicine, the population of the world is soaring. In fig 137, two populations are represented. The first is stable, and change is the result of reproductive success of black variants and failure of the white. In the second example, a similar change occurs, resulting in a generation with three times as many black variants as white (the situation also seen in the first example). However, this second population is not stable, and

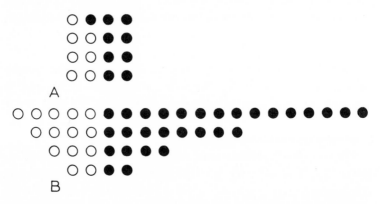

Fig 137 Change in a population can occur for one of two reasons. A variant (black circles) becomes more common as a result of replacement of less successful variants (white circles) in a stable population A, or as a result of having a faster rate of expansion in an unstable population B

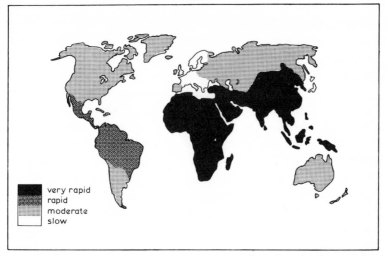

Fig 138 Almost all races of man are expanding, but in some areas expansion is more rapid than in others

the change is the result of the different reproductive rates of black and white variants. That is, both black and white variants are increasing, but the black are increasing at a faster rate.

The situation in fig 137B is a simple model of the human population. With very few exceptions, all races of man are expanding, but some are expanding more rapidly than are others (fig 138). The area of most rapid population growth rates includes Africa, India, and South East Asia, and the areas of fastest growth, in terms of actual increase in numbers, will occur in those parts of this area that are already most densely populated—India and East China. It follows that change in the constitution of the human population will be in favour of the genes and gene combinations that are most common in these parts of the world. What these are, we do not know with precision, but their physical expression is well known. In India the prevailing genes produce a type called the Hindu, and in China, types called Chinese and South East Asian. It is possible that, as a result of economic

217

change, the centres of population explosion will shift. Most likely this shift will be towards Africa, resulting in an increase in the number of negroid types relative to the rest of the world.

There is one last major question concerning man's evolution. Is there any evidence for divergence? The answer appears to be no. Obviously the varied racial types present today are evidence of divergence in the past, but this is unlikely to continue for two reasons. Firstly, a number of races are on the verge of extinction, for example the Murrayian (Australian) and Ainu (East Asia) and, secondly, technological advances make it less and less likely that any population can remain isolated for the length of time needed for isolating mechanisms to evolve.

In conclusion, man is still evolving. There is no evidence for divergent change, but different rates of reproduction in different parts of the world are causing some changes in the overall constitution of his gene pool.

Glossary

abdomen	The hind part of the body in insects and the lower part of the body in vertebrates
acquired character	A character acquired during the life of an individual that is not genetically determined
adaptation	A feature of an organism evolved to perform a particular function. A feature that improves the chances of survival of an organism
adaptive radiation	A number of organisms evolved from a common ancestor that occupy a variety of habitats (figs 126, 127)
adenine	A chemical constituent of DNA (fig 60)
Aeoliscus	Genus to which the shrimp fish or needle fish belongs (fig 104)
albinism	The condition in which pigments are not produced in the body. An animal suffering from this condition is called an albino
allantois	A sac found in the eggs of reptiles, birds, and some mammals into which the embryo excretes waste materials and from which it derives oxygen (fig 13)
alleles	Genes found on similar positions on homologous chromosomes. They have similar functions, but may have different effects. Often, the effects of one of a pair of alleles dominate over the effects of another (fig 66)
amino-acid	An organic chemical constituent of protein
Amphibia	The class to which frogs, newts, and apodans belong
anaphase	A stage in the process of meiotic or mitotic cell division
ancestor	An organism from which another has evolved by simple or divergent change
ancestral type	An organism or hypothetical organism with the characteristics expected in the ancestor to a group of organisms
Anthropoidea	The suborder to which monkeys and man belong

Glossary

articular
One of the bones of the lower jaw in reptiles. Homologous with the malleus in mammals (fig 106)

articulate
Bones that articulate with one another are jointed to one another

artificial selection
The process by which man determines the evolution of an organism by choosing certain offspring to contribute to each successive generation

Australopithecus
The genus to which a possible ancestor of man belongs (fig 132)

Aves
The class to which birds belong

bacterium
A kind of single-celled microscopic organism

barrier
Something that prevents populations of organisms from interbreeding

Beaufort series
A rock formation of Permian and Triassic age that occurs in Africa

biology
The study of life

Biston
Genus to which the peppered moth belongs

bivalent
A pair of homologous chromosomes in close association during meiosis

blastula
A hollow ball of cells, an early stage in the development of vertebrates and other animals (fig 54)

Brookvaliidae
A family of Triassic fossil fish (fig 31)

Camarhynchus
Genus containing the tree finches (fig 25)

Cambrian
A period in the earth's history that began 600 million years ago (fig 86)

Campodea
Genus to which certain dipluran insects belong

Canis
Genus to which the dog belongs (fig 111)

Caransius
Genus to which the stick insect belongs

Carapus
Genus to which the carapus belongs (fig 104)

Carboniferous
A period in the earth's history that began 350 million years ago. In America this period is divided into two parts: the Pennsylvanian and the Mississipian (fig 86)

cartilage
A skeletal tissue with similar functions to bone

Cave sandstone
A rock formation of Triassic age that occurs in Africa

cell
The basic unit of animal and plant tissues. A discrete mass of protoplasm enclosed within a membrane

Cenozoic
An era in the earth's history that began 70 million years ago (fig 86)

Cerion
Genus to which certain land snails belong

Glossary

Charadrius	Genus to which the plover belongs (fig 116)
chitin	A constituent of insect cuticle
Chordata	The phylum to which fish, amphibians, reptiles, birds, and mammals belong
choroid	A layer of cells containing pigment found in the vertebrate eye (fig 12)
chromatid	The two halves of a newly divided chromosome while they are still joined during meiosis and mitosis
chromosome	A structure composed of DNA found in all cells of all organisms. Its function is to organise the production of proteins (fig 58)
Ciconia	Genus to which the stork belongs (fig 49)
class	A taxon used in the classification of animals and plants. A subdivision of a phylum
clavicle	A bone of the shoulder girdle. The collar bone in man (fig 97)
Cleithrolepidina	Genus of Triassic fossil fish (fig 90)
Cleithrolepis	Genus of Triassic fossil fish (fig 90)
Clupea	Genus to which the herring belongs (fig 117)
common ancestor	The ancestor of two or more organisms
continental drift	The process by which the continents have fragmented and moved to their present positions (fig 87)
coracoid	A bone of the shoulder girdle
Craniata	Synonymous with vertebrata
Cretaceous	A period in the earth's history that began 135 million years ago (fig 86)
crossing over	The process by which parts of homologous chromosomes are exchanged during meiosis (fig 72)
Crustacea	The class to which crabs, lobsters, shrimps, and prawns belong
cuticle	A layer of material covering the outer surface of plants and animals. In insects it is composed of chitin
Cyclops	Genus to which certain crustaceans belong
Cynodesmus	Genus of Miocene mammals ancestral to the dog (fig 111)
cytosine	A chemical constituent of DNA (fig 60)
dentary	A bone of the lower jaw in vertebrates. In mammals the only bone of the lower jaw (fig 106)
deoxyribonucleic acid	The substance of which genes are composed. Its molecule consists of two helices of alternating sugar

	and phosphate groups linked by base pairs, adenine and thymine, and guanine and cytosine. The order in which these bases are arranged determines the effects of the gene. Abbreviated to DNA (fig 60)
descendant	An organism that has evolved from another (the ancestor) by simple or divergent change
Devonian	A period in the earth's history that began 400 million years ago (fig 86)
differentiation	The process of change leading to specialisation of cells during development
Dinictis	Genus of Oligocene mammals ancestral to the cat (fig 111)
Diplocaulus	Genus of Carboniferous and Permian amphibians (fig 95)
Diplococcus	Genus to which certain bacteria belong
Diplodocus	Genus of Jurassic dinosaurs
diploid	Having the normal number (two homologous sets) of chromosomes. Abbreviated to *n*
diplotene	A stage in the process of meiotic cell division
divergent change	Change in which an ancestral species gives rise to two or more descendant species (figs 29, 32, 46, 115)
DNA	Abbreviation for deoxyribonucleic acid
dominant	An allele that is effective whether or not it is present in the homozgous or heterozgous condition
Drosophila	Genus to which the fruit fly belongs (fig 50)
Dwyka formation	A rock formation of Permian age that occurs in Africa
ecological barrier	Any factor that separates populations of organisms and prevents interbreeding (figs 118, 121, 124)
ecology	The study of animal and plant communities, their interrelationships and their relationships with their environment
Ectoganus	Genus of Eocene taeniodonts (fig 26)
egg	The female sex cell
embryo	An animal or plant at an early stage of development (fig 14)
endoskeleton	A skeletal system enclosed within the body
environment	The factors that together determine the conditions under which animals and plants live
Eocene	An epoch in the earth's history that began 60 million years ago (fig 86)
epoch	An interval in the earth's history. A subdivision of a period (fig 86)

Glossary

Equus	Genus to which the horse belongs
era	An interval in the earth's history (fig 86)
Escherichia	Genus to which certain bacteria belong
Eusthenopteron	Genus of Devonian lobe finned fish (fig 105)
Exocoetus	Genus to which the flying fish belongs (fig 108)
exoskeleton	A skeletal system that encloses the body
extinction	The fate of a population that leaves no offspring or of a taxon from which no descendant forms evolve
$F_1 F_2 F_3$ etc	Abbreviations for successive generations
family	A taxon used in the classification of animals and plants. A subdivision of an order
family tree	A term synonymous with phylogenetic system
fauna	The community of animals of a particular period of time or place
faunal replacement	The replacement of one fauna by another
Felis	Genus to which the cat belongs (fig 111)
fertilisation	The union of the male and female sex cells to produce a zygote (fig 54)
flora	The community of plants of a particular period of time or place
fossil	The remains of an animal or plant preserved in sedimentary rocks. Also occasionally used with reference to evidence of past life, eg footprints or burrows (fig 38)
fragmentation	The dispersal of a population over two or more separate areas
furcula	The fused clavicles or 'wishbone' of the shoulder girdle in birds (fig 97)
gamete	A sex cell, either a sperm or egg cell. Gametes contain half the normal number of chromosomes
gastrula	A hollow ball of cells with a wall composed of two layers of cells. An early stage in the development of vertebrates and other animals (fig 54)
gene	Part of the DNA molecule responsible for the production of a protein and therefore having a particular effect on the growth and development of an individual
gene combination	A suite of genes responsible for the production of a number of proteins that, together, have a particular effect on the growth and development of an individual

223

Glossary

gene pool	The total collection of genes found in a population of organisms
genotype	The set of genes present in one particular individual
genus	A taxon used in the classification of animals and plants. A subdivision of a family
geological record	The fossil evidence of the past history of plants and animals
geological time scale	The time during which the earth has existed divided into eras, periods, and epochs (fig 46)
Geospiza	Genus to which the ground finches belong (fig 25)
Geospizinae	A subfamily containing Darwin's finches (fig 25)
germ cell	A cell which divides to produce gametes
gill slit	An opening in the throat region. In aquatic animals water passes out of the mouth cavity through the gill slits
Glossopetrae	'Tongue stones'. A name given to fossil shark teeth during the seventeenth century (fig 39)
Glossopteris	Genus of Permian plant (fig 89)
Gryphaea	Genus of Jurassic molluscs (figs 42, 44, 82)
guanine	A chemical constituent of DNA (fig 60)
habitat	The place where conditions are right for a particular animal or plant to live
haemophilia	A disease caused by a recessive gene in men, in which the blood fails to clot
haploid	Having half the normal number of chromosomes. Characteristic of sex cells. Abbreviated to $\frac{1}{2}n$
hemiheterocercal	A type of tail in bony fishes in which a scaly lobe extends about halfway along the upper part of the tail (fig 27)
heterocercal	A type of tail in bony fishes in which there is a complete scaly lobe in the upper part of the tail (fig 27)
heterozygous	The condition in which alleles are different. Aa, Bb, Cc, etc
hexaploid	Having six times the haploid number of chromosomes
hierarchy	An order of groups in which lowly members are small and well defined, and contained within higher groups that are larger and broadly defined (fig 17)
Hominidae	The family to which man belongs (fig 131)
Homo	Genus to which man belongs (fig 133)

Glossary

homocercal	A type of tail in bony fishes in which there is no scaly lobe in the upper part of the tail (fig 27)
Homo diluvii testis	A fossil supposed to be that of a man who was drowned in the Flood (fig 41)
Homoioptera	Genus of Carboniferous insects
homologous	Organs or parts of different organisms are said to be homologous with one another if they are similar and have evolved from a common ancestral organ or part. Homologous chromosomes are chromosomes with the same set of alleles. In all cells the chromosomes occur in homologous pairs, one of each is derived from the mother, one from the father
homozygous	The condition in which alleles are identical, AA, aa, BB etc
humerus	A bone in the fore limb of vertebrates (fig 10)
hybrid	An organism, the result of breeding between members of different populations
hybrid zone	An area between the areas in which different populations occur, in which hybrids are common (fig 120)
Hypoderma	Genus to which certain flies belong (fig 119)
Hyponomenta	Genus to which certain moths belong
Hyracotherium	Genus of Eocene mammals ancestral to the horse
Ichthyosaurus	Genus of Jurassic marine reptiles (fig 98)
Ichthyostega	Genus of Devonian amphibians (fig 105)
incus	One of the bones of the inner ear in mammals. Homologous with the quadrate in reptiles (fig 106)
infertile	Being unable to produce gametes
interbreeding	Breeding between members of different populations
interspecific competition	Competition between different species
intraspecific competition	Competition between members of a single species
inversion	The result of damage and subsequent repair to a chromosome in which a portion of the chromosome is reversed causing a reversal of the order of genes (fig 74)
iris	Part of the vertebrate eye that controls the amount of light entering the eye (fig 12)
isolating mechanism	A mechanism that prevents members of different groups of organisms from interbreeding (fig 122)
Jurassic	A period in the earth's history that began 180 million years ago (fig 86)

Glossary

Kallima   Genus to which the leaf butterfly belongs (fig 103)
keel   A thin sheet of bone attached to the breast plate in birds to which the wing muscles are attached (fig 97)

labium   Part of the mouth in insects. The lower 'lip' (fig 100)
labrum   Part of the mouth in insects. The upper 'lip' (fig 100)
Labyrinthodontia A subclass of the Amphibia containing fossil forms that lived from the Upper Devonian until the Upper Triassic
Lamna   Genus of Eocene shark (fig 39)
Lasiognathus   Genus to which the angler fish belongs (fig 99)
Lepospondyla   A subclass of the Amphibia containing fossil forms common during the Carboniferous and Permian periods
leukaemia   A disease in which too many white blood cells are produced
lineage   A series of phylogenetically related organisms
Lingula   Genus to which certain lamp shells belong
linkage   The association of two or more genes as a result of their occurring on the same chromosome (fig 71)
Lissamphibia   The subclass to which all living amphibians belong
locus   The point of a chromosome at which a particular gene occurs

malleus   One of the bones of the inner ear in mammals. Homologous with the articular in reptiles (fig 106)
mammal   Any member of the class Mammalia
Mammalia   The class to which belong vertebrates with hair, the ability to secrete milk, and a lower jaw composed of a single bone, the dentary
mammal-like reptiles Fossil reptiles, members of which are ancestral to the mammals. They lived from the Middle Carboniferous until the Middle Jurassic
mandible   Part of the mouth in insects. The mandibles are used for crushing and biting food. Also a general word for the lower jaw in vertebrates (fig 100)
marsupial   Any member of the subclass Marsupialia
Marsupialia   The subclass to which pouched mammals belong
maxilla   Part of the mouth in insects. The maxillae lie immediately behind the mandibles. Also, the main tooth-bearing bone of the upper jaw in vertebrates (fig 100)

Glossary

Megaloceros	Genus to which the Irish elk belongs
meiosis	Cell division in which cells are produced, the gametes, which contain half the normal number of chromosomes (fig 69)
Merychippus	Genus of Miocene mammals ancestral to the horse
Mesohippus	Genus of Oligocene mammals ancestral to the horse
Mesozoic	An era in the earth's history that began 225 million years ago (fig 86)
metaphase	A stage in the process of meiotic or mitotic cell division
Miocene	An epoch in the earth's history that began 25 million years ago (fig 86)
Mirabilis	Genus to which the four o'clock flower belongs
mitosis	Cell division in which cells are produced which have the normal number of chromosomes (fig 59)
Monotremata	The subclass to which egg-laying mammals belong
monotreme	Any member of the subclass Monotremata
molecule	A collection of atoms of elements bonded together. The chemical properties of a molecule are determined by its constituents, and its physical properties by its three-dimensional shape
Molteno beds	A rock formation of Triassic age that occurs in Africa
morphology	The form of an organism or of its parts. The study of form
multituberculate	A member of a group of fossil vertebrates that are either mammals or mammal-like reptiles. They lived from Upper Jurassic times until the end of the Eocene (fig 129)
Muridae	The family to which rats and mice belong (fig 129)
mutation	A change in a chromosome (as a result of inversion or translocation), or in a single gene
natural selection	The process by which some members of a population contribute more offspring to subsequent generations than do others
Neoceratodus	Genus to which the Australian lung fish belongs (fig 91)
nerve cord	The nerve that, in vertebrates, runs along the entire length of the body enclosed within the backbone.
notochord	A skeletal rod that runs below the nerve cord in members of the phylum Chordata. It is lost in the adult in most chordates

227

Glossary

Ocypus	Genus to which certain beetles belong
Oligocene	An epoch in the earth's history that began 40 million years ago (fig 86)
order	A taxon used in the classification of animals and plants. A subdivision of a class
Ordovician	A period in the earth's history that began 500 million years ago (fig 86)
organism	An animal, plant, or bacterium
orthogenesis	Evolutionary change that appears to be directed
ossicle	A bone that transmits vibrations in the inner ear. In mammals there are three, the malleus, incus, and stapes
osteoglossid	A member of the Osteoglossiformes, an order of primitive teleost fish.
P	Abbreviation for the parent generation.
Palaeocene	An epoch in the earth's history that began 70 million years ago (fig 86)
Palaeozoic	An era in the earth's history that began 600 million years ago (fig 86)
paleontology	The study of extinct life, especially of evidence provided by fossils
parasite	Any organism dependent upon another for its food and shelter
pectoralis major	The muscle that produces the power for the downstroke of the wing in birds (fig 97)
pectoralis minor	The muscle that produces the power for the upstroke of the wing in birds (fig 97)
period	An interval in the earth's history. A subdivision of an era (fig 86)
Permian	A period in the earth's history that began 270 million years ago (fig 86)
phenotype	The characteristics of an organism, as opposed to the genes that determined them (the genotype)
Phlebotomus	Genus to which certain flies belong
phosphate	A compound of phosphorus and oxygen
Phycodurus	Genus to which the sea dragon belongs (fig 103)
phylogenetic system	A diagram in which the evolutionary relationships of a number of organisms is illustrated (figs 33, 125)
phylogeny	The history of evolutionary change of an organism or group of organisms
Phylum	A taxon used in the classification of animals and plants. A subdivision of a Kingdom
placental	Any member of the subclass Placentalia
Placentalia	The subclass to which belong the mammals that

	retain their young within the mother until a late stage of development
Pleistocene	An epoch in the earth's history that began 4 million years ago (fig 86)
Pliocene	An epoch in the earth's history that began 11 million years ago (fig 86)
Pongidae	The family to which the apes belong (fig 131)
population	A group of individuals that normally only breed amongst themselves. Defined genetically by a gene pool and geographically by a particular area
population sample	A collection of animals or plants in which variants are present in the same proportions as they are present in the population from which the collection was made (figs 43, 51)
Precambrian	The period of the earth's history before Cambrian times
Probainognathus	Genus of Triassic mammal-like reptiles (fig 106)
Probelesodon	Genus of Triassic mammal-like reptiles (fig 106)
prophase	A stage in the process of meiotic or mitotic cell division
protein	A complex organic molecule composed of a number of amino-acids
Prozeuglodon	Genus of Eocene marine mammals
Pseudocynodictis	Genus of Oligocene mammals ancestral to the cats and dogs (fig 111)
Psittacotherium	Genus of Palaeocene taeniodonts (fig 26)
quadrate	One of the bones of the lower jaw in reptiles. Homologous with the incus in mammals (fig 106)
Quaternary	A period in the earth's history that began 4 million years ago (fig 86)
race	A population of individuals with definite characteristics, but able to breed with members of other races. A subdivision of a species
radiation	A number of organisms evolved during a short period from a common ancestor
radius	A bone in the fore limb of vertebrates (fig 10)
Rana	Genus to which the common European frog belongs
recessive	An allele that is not effective unless it is present in the homozygous condition. In the heterozygous condition the effects of the recessive gene are completely masked by those of the dominant gene
Red beds	Rocks of Triassic age that occur in Africa and America. More generally, 'red beds' is a term

	descriptive of any red sedimentary rock usually thought to have been deposited under desert conditions
Redfieldiidae	A family of Triassic fossil fish (fig 31)
Remora	Genus to which the sucking fish belongs (fig 99)
Reptilia	The class to which snakes, lizards, and turtles belong. The class also contains many extinct forms such as the ichthyosaurs, dinosaurs, rhynchosaurs, and pterodactyls
retina	The layer in the vertebrate eye that contains light-sensitive cells (fig 12)
rodent	Any member of the order Rodentia which contains mammals with chisel shaped front teeth such as the rats, mice, guinea pig, and squirrel (fig 129)
Sacculina	Genus to which a crustacean parasite of the crab belongs (fig 15)
Salamandra	Genus to which the salamander belongs (fig 41)
scleroid	The outer layer of the vertebrate eye ball wall (fig 12)
Scleropages	Genus to which primitive teleostean fish from Australia belong (fig 91)
sedimentary rock	Any rock composed of eroded particles of other rock deposited usually in the sea, and in lakes and rivers
segment	A part of the body that is repeated several times within the same animal
segregation	The separation of alleles into different cells during meiotic cell division
selection	See natural selection
sex cell	A gamete, either an egg or sperm. A cell containing half the normal number of chromosomes
Silurian	A period in the earth's history that began 440 million years ago (fig 86)
simple change	Change in which an ancestral species gives rise to a single descendant species (figs 29, 32, 45)
speciation	The evolution of a new species by divergent or simple change
species	A taxon used in the classification of animals and plants. A subdivision of a genus. A species is a population of organisms that breed amongst themselves but not with members of other populations
sperm	The male sex cell
Sphenodon	Genus to which the reptile tuatara belongs (fig 94)
spindle	A structure found in the cell during cell division

stapes	One of the bones of the inner ear in vertebrates (fig 106)
Stylinodon	Genus of Eocene taeniodonts (fig 26)
Stylinodontinae	A subfamily of early Tertiary mammals otherwise called taeniodonts (fig 26)
subclass	A subdivision of a class
subfamily	A subdivision of a family
suborder	A subdivision of an order
subphylum	A subdivision of a phylum
subspecies	A subdivision of a species
taeniodont	Any early Tertiary mammal belonging to the subfamily Stylinodontinae (fig 26)
taxon	Any group used in a classification
taxonomy	The study of animal and plant classification
teleost	Any member of the subclass Teleostei. Almost all bony fishes living today are teleosts
telophase	A stage in the process of meiotic or mitotic cell division
Tertiary	A period in the earth's history that began 70 million years ago (fig 86)
tetraploid	Having four times the haploid number of chromosomes
thorax	The mid part of the body in insects and the upper part of the body in vertebrates
thymine	A chemical constituent of DNA (fig 60)
tissue	An association of cells with the same function, eg, muscle, bone, or cartilage
tongue stone	Glossopetrae. A name given to fossil shark teeth during the seventeenth century (fig 39)
Toxodon	Genus of Pliocene and Pleistocene mammals
Toxotes	Genus to which the archer fish belongs (fig 102)
translocation	The result of damage and subsequent repair of a pair of chromosomes in which portions of each chromosome are exchanged (fig 74)
Triassic	A period in the earth's history that began 225 million years ago (fig 86)
triploid	Having three times the haploid number of chromosomes
ulna	A bone in the fore limb of vertebrates (fig 10)
variant	A member of a population that is different in some respect to other members of that population
variation	Variety in a population of individuals (phenotypic

Glossary

variation) or of gene combinations in a gene pool (genotypic variation)

vertebra
A bone or cartilage of the vertebral column or backbone

Vertebrata
A subphylum of animals characterised by having a skull and backbone (fig 128)

vertebrate
Any member of the subphylum Craniata, a taxon containing the fish, amphibians, reptiles, birds, and mammals

Vulpavus
Genus of Eocene mammals ancestral to cats and dogs (fig 111)

Wortmania
Genus of Palaeocene taeniodonts (fig 26)

yolk sac
A sac found in the eggs of reptiles and birds that contains food for the developing embryo (fig 13)

zygote
A cell, the result of union of the male and female sex cells (fig 54)

Bibliography

Ashley Montague, M. F. (ed). *Culture and the Evolution of Man* (1962)

de Beer, G. *Atlas of Evolution* (1964)

Cain, A. J. *Animal Species and Their Evolution* (1954)

Cannon, H. G. *The Evolution of Living Things* (Manchester 1958)

Colbert, E. H. *Evolution of the Vertebrates* (New York 1955)

Colbert, E. H. *Men and Dinosaurs* (1968)

Darlington, C. D. *The Evolution of Man and Society* (1969)

Dobzhansky, T. *Evolution, Genetics and Man* (New York 1955)

Dobzhansky, T. *Mankind Evolving* (New Haven 1962)

Dobzhansky, T. *Heredity and the Nature of Man* (1965)

Eaton, T. H. *Evolution* (1970)

Ford, E. B. *The Study of Heredity* (Oxford 1950)

Haldane, J. B. S. *The Cause of Evolution* (1932)

Hardy, A. *The Living Stream* (1965)

Hardy, A. *The Divine Flame* (1966)

Howells, W. (ed). *Ideas on Human Evolution* (New York 1967)

Huxley, J. *Evolution, the Modern Synthesis* (1942)

Huxley, J. *Evolution in Action* (1963)

Jepsen, G. L., Simpson, G. G., and Mayr, E. (eds). *Genetics, Paleontology and Evolution* (New York 1963)

Koestler, A. *The Ghost in the Machine* (1970)

233

Maynard Smith, J. *The Theory of Evolution* (1958)

Mayr, E. *Animal Species and Evolution* (Cambridge Mass 1966)

Medawar, P. B. *The Future of Man* (1960)

Monod, J. *Chance and Necessity* (1972)

Olson, E. C. *The Evolution of Life* (1966)

Rook, A. (ed). *The Origins and Growth of Biology* (1964)

Romer, A. S. *Man and the Vertebrates* vols 1 and 2 (1954)

Romer, A. S. *The Vertebrate Body* 3rd edn (1962)

Romer, A. S. *Vertebrate Paleontology* 3rd edn (Chicago 1966)

Simpson, G. G. *The Meaning of Evolution* (New Haven 1949)

Simpson, G. G. *Horses* (New York 1951)

Simpson, G. G. *The Major Features of Evolution* (New York 1953)

Simpson, G. G. *Principles of Animal Taxonomy* (New York 1961)

Simpson, G. G. *This View of Life* (New York 1963)

Simpson, G. G. *The Geography of Evolution* (New York 1965)

Waddington, C. H. *The Nature of Life* (1963)

Wendt, H. *Before the Deluge* (1970)

Young, J. Z. *The Life of Vertebrates* (Oxford 1950)

Young, J. Z. *An Introduction to the Study of Man* (Oxford 1971)

Index

Figures in italic denote illustrations

235

Index

Index

tools of man, *206, 211*
Toxodon, 24
Toxotes, 158
translocation, *118*, 231
trout, isolation of, 145, *146*

variants, their spread in a
population, 76, *77*, 78, 231

variation, caused by the
environment, 78, *79*, 231
vertebrate skeleton, *18, 22, 23,
24*
vertebrates, history of, *200*
Vulparus, 173

Wortmania, 46, *47*